Human Genetic Disease Analysis

A Practical Approach

Second edition

Edited by

K. E. DAVIES

Molecular Genetics Group, Institute of Molecular Medicine,
John Radcliffe Hospital, Oxford OX3 9DU, UK

OXFORD UNIVERSITY PRESS
Oxford New York Tokyo

Oxford University Press, Walton Street, Oxford OX2 6DP
Oxford   New York   Toronto
Delhi   Bombay   Calcutta   Madras   Karachi
Kuala Lumpur   Singapore   Hong Kong   Tokyo
Nairobi   Dar es Salaam   Cape Town
Melbourne   Auckland   Madrid
and associated companies in
Berlin   Ibadan

Oxford is a trade mark of Oxford University Press

A Practical Approach 🔵 is a registered trade mark
of the Chancellor, Masters, and Scholars of the University of Oxford
trading as Oxford University Press

Published in the United States
by Oxford University Press Inc., New York

A catalogue record for this book is available from the British Library

Library of Congress Cataloging in Publication Data
Human genetic disease analysis : a practical approach/edited by K. E.
Davies. — 2nd ed.
(Practical approach series)
Rev. ed. of: Human genetic diseases. c1986.
Includes bibliographical references and index.
1. Genetic disorders—Diagnosis.   I. Davies, K. E. (Kay E.)
II. Human genetic diseases.   III. Series.
[DNLM: 1. DNA—diagnostic use.   2. Hereditary Diseases—diagnosis.
QZ 50 H9165]
RB155.6.H85 1993   616'.042—dc20   92–48251
ISBN 0–19–963308–8 (p/b)
ISBN 0–19–963309–6 (h/b)

Typeset by Footnote Graphics, Warminster, Wilts
Printed in Great Britain by Information Press Ltd, Eynsham, Oxon

Preface

The first edition of this book was written at a time when DNA recombinant technology was being introduced into diagnostic and research laboratories. The field has advanced very rapidly since then with major centres providing diagnosis for a whole range of important genetic disorders based on genomic analyses. The advent of the polymerase chain reaction (PCR) has had a major impact on all relevant techniques. For example, prenatal diagnosis which used to take two to three weeks is now being performed within twenty four hours of a sample being taken. Fluorescence *in situ* hybridization (FISH) techniques have been optimized such that markers can readily be ordered between 100 kb and 2–3 megabases apart. Thus the gap between the genetic linkage marker and the disease locus has been bridged. In addition FISH is proving to be a very powerful tool in its own right for the analysis of chromosomal rearrangements.

This edition covers many of the major approaches to gene analysis both in the long range investigation using pulsed field gel electrophoresis or FISH, and in the detailed analysis of the transcriptional unit of a gene. I am very grateful to the authors for their hard work in producing their excellent up-to-date contributions. I hope that the book will serve as a valuable guide for anyone wishing to analyse a particular genetic disease whether for pure research purposes or for genetic counselling.

Oxford   K.E.D.
November 1992

Contents

3. Pulsed-field gel electrophoresis in the analysis of genomic DNA and YAC clones 35

J. T. den Dunnen, P. M. Grootscholten, and
G. J. B. van Ommen

4. Fluorescent *in situ* hybridization 59

V. J. Buckle and K. A. Rack

Contents

5. Fine mapping of genes: the characterization of the transcriptional unit   83

M. Antoniou, E. deBoer, and F. Grosveld

Contents

6. Chromosome analysis and sorting by flow cytometry

S. Monard, L. Kearney, and B. D. Young

Contributors

M. ANTONIOU
Gene Structure and Expression, National Institute for Medical Research, The Ridgeway, Mill Hill, London NW7 1AA, UK.

V. J. BUCKLE
MRC Molecular Haematology Unit, Institute of Molecular Medicine, John Radcliffe Hospital, Headington, Oxford OX3 9DU, UK.

E. DEBOER
Gene Structure and Expression, National Institute for Medical Research, The Ridgeway, Mill Hill, London NW7 1AA, UK.

J. T. DEN DUNNEN
Department of Human Genetics, Sylvius Laboratory, Leiden University, Wassenaarseweg 72, 2333 AL Leiden, The Netherlands.

ALI EHSANI
Division of Biology, Beckman Research Institute of the City of Hope, Duarte, California 91010-0269, USA.

P. M. GROOTSCHOLTEN
Department of Human Genetics, Sylvius Laboratory, Leiden University, Wassenaarseweg 72, 2333 AL Leiden, The Netherlands.

F. GROSVELD
Gene Structure and Expression, National Institute for Medical Research, The Ridgeway, Mill Hill, London NW7 1AA, UK.

L. KEARNEY
ICRF Medical Oncology Department, St Bartholomew's Hospital, London EC1, UK.

S. MONARD
ICRF, Lincoln's Inn Fields, London, UK.

JOHN M. OLD
National Haemoglobinopathy Reference Service, Institute of Molecular Medicine, John Radcliffe Hospital, Oxford OX3 9DU, UK.

K. A. RACK
MRC Molecular Haematology Unit, Institute of Molecular Medicine, John Radcliffe Hospital, Headington, Oxford OX3 9DU, UK.

Contributors

SWEE LAY THEIN
MRC Molecular Haematology Unit, Institute of Molecular Medicine, John Radcliffe Hospital, Headington, Oxford OX3 9DU, UK.

G. J. B. VAN OMMEN
Department of Human Genetics, Sylvius Laboratory, Leiden University, Wassenaarseweg 72, 2333 AL Leiden, The Netherlands.

R. BRUCE WALLACE
Division of Biology, Beckman Research Institute of the City of Hope, Duarte, California 91010-0269, USA.

B. D. YOUNG
ICRF Medical Oncology Department, St Bartholomew's Hospital, London EC1, UK.

Abbreviations

ARMS   amplification refractory mutation system
ASO   allele-specific oligonucleotide
BSA   bovine serum albumin
BUdR   5-bromodeoxyuridine
cDNA   complementary DNA
CHEF   contour clamp homogeneous electric field
CVS   chorionic villus sample
DAPI   4′,6-diamidino-2-phenylinodole
DIG   digoxigenin
DMSO   dimethylsulphoxide
DNA   deoxyribonucleic acid
DOP-PCR   degenerate oligonucleotide primed polymerase chain reaction
DTT   dithiothreitol
EDTA   ethylenediamine tetraacetic acid
EM   electron microscopy
FIGE   field inversions gel electrophoresis
FISH   fluorescence *in situ* hybridization
FITC   fluorescence isothiocyanate
Hepes   N-2-hydroxyethylpiperazine-N′-2-ethanesulphonic acid
HSR   homogenously staining chromosomal regions
Mb   million base pairs
MOPS   3-N-morpholino-propane sulphonic acid
NFDM   non-fat dried milk
NP-40   Nonidet P-40
PBS   phosphate-buffered saline
PCR   polymerase chain reaction
PEG   polyethylene glycol
PFG   pulsed field gradient
PFGE   pulsed-field gel electrophoresis
PHA   phytohaemagglutinin
PI   propidium iodide
Pipes   piperazine-N,N′-bis-2-ethane sulphonic acid
PMSF   phenylmethylsulphonylfluoride
RFE   rotating field electrophoresis
RFLP   restriction fragment length polymorphism
RNA   ribonucleic acid
SDS   sodium dodecyl sulphate
SET   saline-EDTA–Tris
SSC   standard saline citrate

Abbreviations

TAFE	transverse alternating field electrophoresis
Td	dissociation temperature
TE	Tris–EDTA
Tm	melting temperature
Tr	reassociation temperature
YAC	yeast artificial chromosome
YPD	yeast extract, pepton, dextrose

1

Fetal DNA analysis

JOHN M. OLD

1. Introduction

The application of molecular biology techniques to the study of genetic disease has provided a wealth of information about their molecular basis, and this knowledge has been used to establish comprehensive prenatal diagnosis programmes by fetal DNA analysis. The first prenatal diagnosis was for α^{o}-thalassaemia in 1976 using a simple hybridization technique of an α-globin gene probe to amniocyte DNA. The discovery of restriction fragment length polymorphisms enabled sickle cell anaemia and β-thalassaemia to be diagnosed in amniocyte DNA, but it was not until 1982 with the development of chorionic villus sampling (CVS) that prenatal diagnosis by fetal DNA analysis became widely established.

The application of allele-specific oligonucleotide hybridization to detect point mutations started a movement away from indirect analysis to the direct detection of mutations and the switch over became complete after the discovery of the polymerase chain reaction (PCR) in 1985 (1). PCR made prenatal diagnosis quicker and easier to carry out with very small quantities of DNA. It has also inspired the development of new approaches to the direct detection of mutations in fetal DNA, such as the amplification refractory mutation system (ARMS), and has opened up new approaches to fetal DNA analysis such as the analysis of fetal cells in maternal blood, and the diagnosis of mutations in embryonic cells or polar bodies.

2. Chorionic villi DNA

The major advantages of a first trimester diagnosis are that it reduces the stressful period of waiting for the results and, if the fetus is affected, the couple can opt for a termination of the pregnancy when the risks to the mother are still minimal. Chorionic villi can be used for karyotyping, chromosome analysis, biochemical tests for the detection of many inherited metabolic disorders, and as a source of DNA for gene analysis. Thus although one or two questions still remain about the accuracy and safety of chorionic villus sampling (1), it has rapidly become the most important method for prenatal diagnosis.

2.1 Sampling and shipment

Chorionic villus sampling is normally carried out between 8 and 13 weeks gestation by either a transcervical or a transabdominal technique, although the latter approach may be used later in pregnancy as an alternative to amniocentesis or fetal blood sampling for source of fetal DNA. Villi collected by either method must be sorted to remove all traces of material tissue before shipment or DNA preparation. This is usually done by the centre performing the CVS or by arrangement at a local cytogenetics laboratory. The villi should be flushed into a Petri dish for examination under a low-power dissecting microscope and any contaminating maternal cells teased apart and removed. The villi can then be sent to the DNA analysis laboratory for prenatal diagnosis.

Sorted villi may be sent in a small amount of tissue culture medium at room temperature if the shipment time is 24 hours or less. However if a longer time in transit is anticipated it is better to send the sample at room temperature either in lysis buffer, or if the CVS centre can prepare DNA, as purified DNA. Lysis buffer (10 mM NaCl, 25 mM EDTA, 0.1% SDS, 50 µg/ml proteinase K) can be prepared by the DNA laboratory and sent to the CVS centre in advance. The sorted villi should be pelleted by centrifugation in a 1.5 ml Eppendorf tube and resuspended in 0.5 ml of the lysis buffer. The villi slowly lyse in transit and the sample can be incubated as normal upon arrival to complete the lysis. The DNA is quite stable in lysis buffer for several weeks as we found out by accident when such a sample was misplaced by an air courier service and was delivered 14 days after the shipping date but without any ill effect to the DNA (a diagnosis was obtained).

2.2 DNA preparation

The following protocol is based on a scaled-down version of our published protocol for the extraction of DNA from peripheral blood samples (2). If the sample is sent in lysis buffer, proceed straight to step **4**. When extracting DNA from more than one sample at the same time great care must be taken not to mix up or cross contaminate the samples. We find it most convenient to use colour coded Eppendorf tubes in such instances.

Protocol 1. Preparation of DNA from chorionic villi

1. Transfer sorted villi to a 1.5 ml Eppendorf tube. Centrifuge for 30 sec and remove tissue culture medium.

2. Wash villi by resuspension in 0.5 ml of 150 mM NaCl, 25 mM EDTA. Centrifuge for 30 sec and remove aqueous layer.

3. Add 0.5 ml of lysis buffer (150 mM NaCl, 25 mM EDTA,[a] 0.1% SDS, 50 μg/ml proteinase K) and vortex.

4. Incubate at 55°C for 3 h, or overnight at 37°C.

5. Add 0.25 ml of chloroform and 0.25 ml of phenol.[b] Close centrifuge tube cap tightly and vortex mixture for 30 sec.

6. Centrifuge for 1 min and transfer the upper aqueous layer to a clean Eppendorf tube.

7. Carry out a second phenol extraction by repeating steps 5 and 6.

8. Add 0.5 ml of chloroform and mix thoroughly.

9. Centrifuge for 1 min and transfer the upper aqueous layer to a clean Eppendorf tube.

10. Repeat step 8.

11. Centrifuge for 1 min and transfer the upper aqueous layer to a 5 ml plastic tube.

12. Add 250 μl of 7.5 M ammonium acetate and mix thoroughly. Add 1.5 ml of absolute ethanol and mix thoroughly. At this stage the DNA is usually observed to precipitate as a clump of white fibrous strands.

13. If the DNA is not observed to precipitate, proceed to step 16. Otherwise simply pellet the DNA by centrifugation for 1 min and remove the ethanol (which still contains the RNA in suspension).

14. Rinse the DNA pellet with 70% ethanol and leave tube inverted to air-dry the pellet at room temperature for 15 min.

15. Redissolve DNA pellet in 50 μl of distilled water (or 25 μl or 100 μl if the amount of DNA is very small or very large).

16. If the DNA concentration is too dilute to precipitate at step 12, or if the maximum yield of DNA is required, the ethanol mixture should be cooled by either placing it at −20°C for at least 1 h or by submerging the tube in dry-ice pellets for 15 min before centrifugation at maximum speed for 20 min. The resulting pellet will comprise of both DNA and RNA, and is treated accordingly by either:

    (a) If the presence of RNA does not matter, process the pellet by steps 14 and 15. (RNA does not interfere with restriction enzyme digestion or DNA amplification. However an accurate estimation of DNA concentration cannot be made.)

    (b) If removal of the RNA is required, the pellet can be dissolved and treated with RNase, previously boiled to remove any DNase activity. However we prefer to simply dissolve the pellet in 50 μl distilled water to concentrate the DNA, and then precipitate by adding 5 μl of 4 M NaCl and 110 μl absolute ethanol at room temperature. If

Protocol 1. *Continued*

      DNA is observed, repeat steps **14** and **15**. If not, repeat step **16**
      before **14** and **15**, and redissolve pellet in 25 µl distilled H$_2$O.

  [a] Stock solution of 500 mM EDTA is prepared by adding 18.6 g to 100 ml dH$_2$O. Adjust pH to
7.5 by adding NaOH pellets until EDTA just dissolves.
  [b] Use fresh ultrapure or redistilled grade phenol (discard if crystals are pink). Dissolve 1.5 kg
in 165 ml dH$_2$O. Add 200 ml of 1 M Tris–HCl (pH 8.0), mix, and remove aqueous layer. Repeat
twice. Add 8-hydroxyquinolene to a final concentration of 0.1%. Add an equal volume of 0.1 M
Tris–HCl (pH 8.0), 0.2% β-mercaptoethanol. Remove aqueous layer and store in covered
bottle. Phenol is an extremely dangerous chemical and should be handled with the greatest of
care.

2.3 DNA yield

Chorionic villi are an excellent source of DNA and a yield of at least 1 µg
DNA per milligram wet weight of villi should be obtained by following
Protocol 1. The size of the villus biopsy can vary enormously from as little as 5
mg to up to 100 mg or more, and the yield of DNA varies in proportion
(*Figure 1*). The extraction procedure can be relied upon to prepare DNA
from very small samples of under 5 mg, but in our experience the average
yield of DNA per sample is around 35 µg, more than sufficient for Southern
blot analysis. We normally digest 5 µg samples for Southern blots, and
amplify approximately 0.5 µg per PCR analysis.

    The DNA concentration can be measured by pipetting a 2.5 µl aliquot into

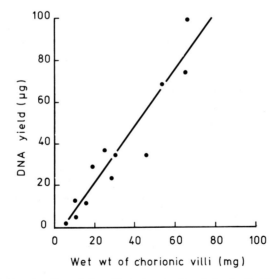

Figure 1 Graph showing the relationship between size of chorionic villus sample and the
yield of DNA.

a quartz semi-micro cuvette containing 1 ml distilled water, mixing the solution by inverting several times, and then reading the absorbance at 260 nm. A DNA solution of 1 μg/ml will absorb 0.020 optical density units. Thus for example if a 2.5 μl aliquot from 50 μl of DNA solution gives a reading of 0.040, this would represent a DNA concentration of 2 μg/ml in the cuvette, and a yield of 40 μg.

The purity of the DNA can be checked by measuring the absorbance at both 260 nm and 280 nm. The 260/280 ratio should be close to a value of 1.8. Samples giving a figure outside the range of 1.6–2.0 contain impurities such as protein or phenol and should be re-extracted.

3. Amniocyte DNA

Amniocentesis is generally accepted as the invasive procedure for obtaining fetal cells that carries the least risk to the fetus. It is usually performed between 16 and 18 weeks gestation, although recently the analysis of amniotic fluid DNA obtained in the first trimester between 8 and 14 weeks has been described (3). Amniocentesis is often used to obtain fetal DNA for couples who present themselves too late for chorionic villus sampling, or for couples who require extensive family studies before the feasibility of prenatal diagnosis can be assessed.

3.1 Choice of sample

DNA may be prepared from amniotic fluid directly or from cell cultures. The disadvantage of using cultured amniocytes is that it takes two to three weeks to grow the cells to confluency in a 25 ml flask, and therefore the diagnosis is prolonged by this period. The advantage is that large amounts of fetal DNA are reliably obtained from cultured cells. Usually two or three flasks are set up for culture, and the yield of DNA from one flask of confluent cells is around 25 μg of DNA, sufficient for all purposes.

The disadvantages of preparing DNA directly from amniotic fluid samples are that small yields of DNA are obtained, and that there is a considerable failure rate to obtain a diagnosis either through insufficient DNA, or poor quality DNA which fails to amplify or give bands on a Southern blot. In our experience the failure rate has been 11 out of 54 diagnoses by Southern blotting (20%), and 3 out of 24 diagnoses by PCR (12%). In contrast we have had no failures in 15 diagnoses using cultured amniocyte DNA. The yield of DNA from 20 ml of amniotic fluid varies between 1.5 μg and 7.0 μg and is just enough for one Southern blot analysis.

The best approach is for 30 ml of amniotic fluid to be taken and 10 ml used for karyotyping and establishing cell cultures, while the remaining 20 ml is sent to the DNA diagnosis laboratory for immediate analysis. Amniotic fluid can be sent by 24 hour delivery at room temperature. However for longer shipment times it is best sent as a frozen amniocyte pellet packed in dry-ice, or as a cell pellet resuspended in 0.1 ml of lysis buffer (note the smaller volume compared to chorionic villi because of the smaller amount of DNA present).

3.2 DNA preparation

The method of DNA preparation is essentially the same as that described for chorionic villi.

Protocol 2. Preparation of DNA from amniocytes

1. Suspend frozen amniocyte pellet in 0.1 ml of lysis buffer (see *Protocol 1*). For amniotic fluid, centrifuge cells, remove supernatant, and add 0.1 ml of lysis buffer. Cultured cells are usually transported at room temperature in flasks topped up with medium. Decant most of the medium and scrape cells off the flask surface into the remaining medium. Transfer to centrifuge tube and pellet cells by centrifugation. Remove supernatant and add 0.1 ml of lysis buffer. Transfer to 1.5 ml Eppendorf tube.

2. Incubate at 55°C for 3 h or 37°C overnight.

3. Add 50 μl chloroform and 50 μl phenol. Mix and spin.

4. Remove aqueous layer to new tube and repeat step **3**.

5. Add 100 μl chloroform, mix, and spin.

6. Remove aqueous layer to new tube and repeat step **5**.

7. Remove aqueous layer to new tube, add 50 μl of 7.5 M ammonium acetate, and 300 μl of absolute ethanol. If a fibrous DNA precipitate is observed, collect by centrifugation, wash in 70% ethanol, and dissolve the air-dried pellet in 25 μl distilled water as described in *Protocol 1*.

8. If no DNA precipitation occurs, cool the mixture at −20°C for 1 h, or for 15 min in dry-ice, and centrifuge for 20 min at maximum speed.

   A tiny pellet of RNA plus DNA is usually obtained which can be carefully rinsed in 70% ethanol, air-dried, and redissolved in 25 μl of distilled water. It is not worth trying to separate the DNA from the RNA by reprecipitation as described for chorionic villi DNA as what little amniocyte DNA there is may be lost in the process.

---

4. Other DNA sources

We have occasionally used fetal blood as a source of fetal DNA in order to confirm a diagnosis obtained by globin chain synthesis. Although for the prenatal diagnosis of thalassaemia CVS DNA analysis has replaced fetal blood sampling and globin chain synthesis in most centres, it is still used occasionally for couples presenting between 18–20 weeks gestation where the mutations have not been previously studied or are unknown. Current fetal blood sampling techniques yield pure fetal blood and 25 μg of DNA can be obtained from a 0.5 ml aliquot. The blood sample is lysed with 1 ml of 155 mM NH$_4$Cl, 10 mM KHCO$_3$, 0.1 mM EDTA in a 1.5 ml Eppendorf tube, and a nuclear pellet collected by centrifugation. The lysate is removed and the fetal DNA is extracted from the nuclear pellet using *Protocol 1*.

Fetal nucleated erythrocytes circulating in maternal peripheral blood in the first trimester of pregnancy are also a source of fetal DNA, and they offer an attractive non-invasive approach to fetal diagnosis for X-linked disorders and paternally inherited mutations. Fetal DNA from these cells has been detected by the amplification of a Y chromosome specific sequence in fetal cells sorted by flow cytometry from maternal blood samples taken between 11 and 16 weeks gestation (4). A Y chromosome sequence has also been detected by amplification with nested primers in unsorted maternal blood samples (5). However until 100% pure fetal cells can be separated the contamination with maternal cells will prevent the complete prenatal diagnosis of autosomal recessive disorders such as thalassaemia and cystic fibrosis by this approach.

5. Quick DNA preparation for PCR

Because the PCR technique is so sensitive amplification of DNA can be carried out directly from tissue samples without the need for DNA extraction. The requirements are for the cell and nuclear membrane to be lysed, and for the DNA to be denatured. Both can be achieved by boiling, and successful amplification from whole blood has been reported by simply adding 1 μl of blood to 100 μl standard PCR mixture, and performing three cycles of 94°C–3 min/55°C–3 min instead of the initial denaturation step (6). However it is probably better to prepare a lysate mixture which can be used for several amplification reactions. Because the DNA is unpurified this quick DNA approach is prone to inhibitory effects. For example the addition of 4 μl of whole blood instead of 1 μl as described above is reported to totally inhibit the reaction, and also likewise if too much lysate is used. *Taq* polymerase is also sensitive to the presence of detergents, especially SDS, so only mild ones such as NP-40 should be used. Lysates may be prepared from chorionic villi or amniocytes as described in *Protocol 3*, although in our laboratory we always use DNA extracted by the phenol–chloroform method.

Protocol 3. Quick DNA preparation for PCR

1. Pellet chorionic villi or amniocytes from 1 ml of amniotic fluid by centrifugation.

2. Remove supernatant and add 100 μl of 1% NP-40.

Protocol 3. *Continued*

3. Vortex or pipette up and down.
4. Add 100 μl of light paraffin oil.
5. Incubate at 100°C for 10 min.
6. Centrifuge for 5 min to pellet cell debris.
7. Use supernatant for PCR. (A 1–5 μl aliquot for a 25 μl PCR mixture should be sufficient.)

6. Storage of DNA

DNA solutions may be stored at 4°C, at −20°C, or at −70°C. Keeping DNA samples at 4°C means that it does not undergo repetitive freeze-thaw cycles while it is being used. It has been proposed that repeated freezing and thawing may damage high molecular weight DNA by shearing, although in our experience it does not appear to harm the DNA for the purpose of restriction endonuclease analysis. However in our laboratory we find it convenient to keep the DNA samples at 4°C for the duration of the diagnostic tests, and then to transfer them to −20°C for long-term storage.

Our DNA bank consists of upright domestic freezers containing a sectioned aluminium tray storage system commissioned by Denley Instruments Ltd. to fit on each of the shelves. Each tray contains 200 numbered 1.5 ml Eppendorf tubes, giving a total storage capacity for 5000 samples per freezer. The appropriate tray can be removed and a particular sample located within seconds. In our experience this is a more user-friendly storage system than a similar set up in a −70°C freezer, and we have kept purified DNA samples at −20°C for ten years to date without any deterioration. However for long-term storage of tissue samples it is generally agreed that −70°C is necessary to prevent DNA degradation.

7. DNA analysis

Mutations and gene deletions in fetal DNA can be detected directly by either Southern blot analysis, or by one or more of the PCR-based techniques. A diagnosis by PCR methods can take as little as four hours whereas Southern blotting still takes a minimum of five days in our laboratory. Thus once a gene defect has been characterized a PCR-based diagnosis technique becomes the method of choice. Cystic fibrosis, Duchenne muscular dystrophy, sickle cell anaemia, and thalassaemia are all diagnosed by PCR methods now in most laboratories. Genetic disorders for which the gene defect is unknown, such as Huntington's disease, myotonic dystrophy, adult polycystic kidney disease, are still diagnosed by Southern blot analysis using DNA probes for flanking marker polymorphisms, although even in these cases, once the marker sequences have been characterized a PCR-based approach is preferred.

It is beyond the scope of this chapter to describe detailed protocols for each of the different techniques for fetal DNA analysis. There are many laboratory manuals and up-to-date texts containing in-depth practical details. In particular the manual by T. Maniatis and co-workers (7) is recommended for the basic techniques such as DNA preparation, restriction enzyme analysis, and Southern blotting, while for the PCR-based approaches there are two recently published guides (8, 9).

7.1 Southern blot analysis

This widely-used technique was first introduced by Southern in 1975 and has since been developed and modified extensively by advances in technology. The initial method described the detection of gel-fractionated DNA fragments following transfer to a nitrocellulose membrane. DNA (5 or 10 μg) is digested by a restriction endonuclease and subjected to electrophoresis in a horizontal agarose gel. After running the gel for sufficient time to separate the fragments, the gel is stained with ethidium bromide and photographed under ultraviolet illumination to provide a visual record. The gel is then soaked in alkali to denature the DNA, neutralized, and set up for transfer of the fragments on to a nitrocellulose filter by a process of blotting the buffer through the gel and filter. The filter is baked to fix the DNA, and then hybridized to a DNA probe labelled by nick translation or hexanucleotide priming with [^{32}P]dCTP after a short pre-soaking to reduce the background signal. The filter is washed to remove unhybridized probe and subjected to autoradiography with X-ray film to visualize the hybridized DNA fragments.

The recent advances include the use of nylon or compound nylon-nitrocellulose membranes instead of nitrocellulose, to allow repeated probings and quicker methods of DNA fixing. The transfer process can be shortened by using a specialized vacuum blotting apparatus although only one gel can be blotted at a time, and for multiple blots the home-made paper towel set-up shown in *Figure 2* is still the most convenient method. The hybridization and washing steps can now be optimized by using glass bottles in a specialized oven containing a rotisserie device. This is a great improvement upon the old plastic bag technology, giving better results and safer handling of radio-activity. Various non-radioactive labelling kits are now commercially available but the use of ^{32}P-labelled probes remains the most popular technique.

7.2 Polymerase chain reaction

The polymerase chain reaction amplifies enzymatically a short specific DNA sequence through repeated cycles of heat denaturation of the DNA, annealing of two oligonucleotide primers that flank the DNA fragment, and extension of the annealed primers with a heat-stable DNA polymerase (10). The process results in an exponential synthesis of the target fragment from

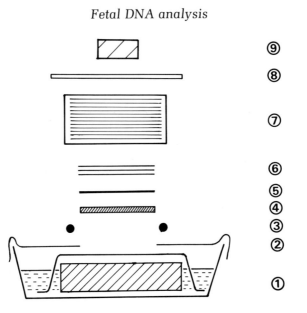

Figure 2 Cross section of a simple Southern blot apparatus. The components are: (1) tray containing 20 × SSC and a slab of saturated foam sponge covered with a sheet of thick filter paper; (2) cling-film cover cut to surround the gel; (3) support for paper towels, e.g. a 10 ml pipette; (4) one or more agarose gels; (5) nitrocellulose or nylon filter; (6) three sheets of filter paper cut to gel size; (7) stack of paper towels; (8) glass sheet; (9) weight.

the third cycle so that after 30 cycles approximately ten million (2^{27}) copies of the target fragment will have been produced. The amount of amplified product is sufficient to be visualized by ethidium bromide staining following restriction enzyme digestion, or more than enough to be hybridized with a labelled allele-specific oligonucleotide probe in a simple dot-blot, thus making the analysis of single gene sequences much quicker and simpler to perform.

7.2.1 Standard PCR conditions

The standard conditions recommended by Cetus for use with their cloned *Taq* polymerase (Amplitaq) will amplify most target sequences. However some primers may need a different annealing temperature or higher $MgCl_2$ concentration for optimum amplification. The standard conditions are a 100 μl reaction volume containing 50 mM KCl, 10 mM Tris–HCl (pH 8.3), 1.5 mM $MgCl_2$, 100 μg/ml gelatin, 0.2 μM of each primer, 200 μM of each dNTP, 1 μg of DNA, 2.5 U of *Taq* polymerase, and a 50 μl overlay of light paraffin oil. Usually 25 to 35 cycles are performed with a profile of 94°C for 1 min, 55°C for 1 min, and 72°C for 1.5 min. A final extension step at 72°C for 3 min usually concludes the cycling. For target sequences longer than 1 kb the extension step should be lengthened by 1 min/kb.

In our laboratory we use a quarter of the above specified amounts by reducing the reaction volume to 25 μl for economy reasons. As few pipettings as possible are done by making up pre-mixes. We have found that an initial denaturation step does not seem to be necessary and the type of PCR machine used appears to make little difference to the result. Our methodology is described below in *Protocol 4*.

Protocol 4. A method for DNA amplification

1. Prepare a stock of 10 × buffer. Add 0.5 ml 1 M Tris–HCl (pH 8.3), 1.23 ml 2 M KCl, 75 μl 1 M MgCl$_2$, 5 mg gelatin to 3.275 ml distilled water. Warm to dissolve the gelatin.

2. Prepare a stock dNTP mixture (1.25 mM each dNTP). Add 50 μl each of 100 mM dATP, dCTP, dGTP, and dTTP to 3.8 ml distilled water. We use commercially made dNTP solutions.

3. Prepare a reaction mixture by adding 0.5 ml of 10 × buffer, 0.8 ml of 1.25 mM dNTP mixture to 2.7 ml of distilled water.

4. Pipette 20 μl of reaction mixture into a 0.5 ml tube.

5. Add 1 μl of each primer (at 100 U/ml).

6. Add 0.5 U of *Taq* polymerase. This is supplied at 5 U/μl and it is easiest to make a small dilute working solution of 0.5 U/μl.

7. Add 1 μl of DNA.

8. Add 25 μl of light paraffin oil.

9. Place in PCR machine set, for example, at 97°C–1 min/55°C–1 min/ 72°C–1.5 min for 30 cycles, plus final step of 72°C–3 min.

10. Remove an aliquot for gel electrophoresis, dot-blot analysis, restriction enzyme digestion etc.

7.2.2 Gel electrophoresis

Small gene deletions and DNA polymorphisms due to different lengths of repeated DNA sequence can be identified by simply running the amplified product in an agarose or polyacrylamide minigel and visualizing the differently sized fragments by ethidium bromide staining and ultraviolet light illumination. For example the 619 bp deletion in the β-globin gene responsible for one type of thalassaemia in Asian Indians can be detected by designing flanking primers that amplify across the deletion break points. Such primers (see *Figure 3*) produce a characteristic product which is 619 bp shorter than that produced by normal DNA. Thus the deletion can be detected in heterozygotes as well as individual homozygotes for the mutation.

Large gene deletions where the breakpoint sequences are unknown are detected by the absence of any amplified product from the undetected normal DNA sequence. Obviously carriers can not be identified by this approach. More than one deletion gene can be analysed for at the same time. A method

Fetal DNA analysis

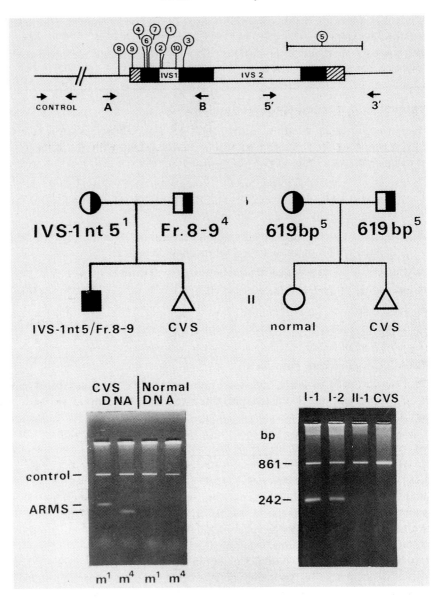

Figure 3 Prenatal diagnosis of β-thalassaemia. A map of the β-globin gene shows the position of the ten most common Asian Indian β-thalassaemia mutations. The 619 bp deletion (no. 5) is detected by amplification with flanking primers (5′ and 3′). A diagnosis of a normal fetus is shown on the *right*. The remaining mutations are detected using the ARMS technique, as illustrated for IVS-1 nucleotide 5 (G-C) (no. 1), and frameshift 8–9 (+G) (no. 4), by amplifying with two control primers and an ARMS primer to detect the mutation (m^1 or m^4) coupled with either primer A or B. A diagnosis of an affected fetus (a compound heterozygote for both mutations) is shown on the *left*.

of screening for Duchenne muscular dystrophy deletions in male DNA by the simultaneous analysis of six frequently deleted dystrophin exons was first introduced by Chamberlain (11). Multiplex PCR amplification has subsequently been developed to the point where 98% of all dystrophin deletions detectable by Southern blot analysis can now be screened for (12).

7.2.3 Restriction enzyme digestion

Point mutations which either create or abolish a restriction endonuclease site can be analysed by digesting the amplified product with the appropriate restriction enzyme before subjecting the fragments to gel electrophoresis. Sickle cell anaemia results from a point mutation which destroys a recognition site for the enzyme *Dde*I. Thus the sickle cell mutation can be screened for by amplification of part of the β-globin gene followed by *Dde*I digestion (*Figure 4*). Note that the amplified fragment contains several other flanking *Dde*I sites which act as internal controls for complete digestion of the amplified product.

Restriction fragment length polymorphisms can be analysed in a similar manner if the DNA sequence surrounding the polymorphic site is known. This makes haplotyping extremely quick and simple and we routinely use linkage analysis of β-globin RFLPs to confirm a diagnosis of β-thalassaemia made by direct detection of mutations whenever possible.

7.2.4 Dot-blot analysis

Point mutations and small nucleotide insertions or deletions can be analysed by the hybridization of labelled allele-specific oligonucleotide probes. The amplified product is denatured and blotted on to a nylon filter using a vacuum dot-blot apparatus. The DNA is hybridized to the labelled probe specific to the mutation being screened for, and then washed either in tetramethylammonium chloride, or at a specific temperature calculated from the T_m of the probe, in order to remove the probe from mismatched hybrids. For prenatal diagnosis fetal DNA must be hybridized separately to a mutant probe and a normal probe complementary to the DNA sequence at the site of the mutation. This approach is not very suitable for screening a DNA sample for many different possible mutations and therefore the reverse approach is under development. This involves immobilizing the oligonucleotide probes for all the possible mutations as individual dots on one filter and then hybridizing the target DNA sequence after amplification and labelling.

7.2.5 Allele-specific priming

Point mutations can also be diagnosed directly by the presence or absence of amplification with allele-specific primers. This method, called the amplification refractory mutation system (ARMS), was introduced for the analysis of α1-antitrypsin alleles (13) and has since been developed for the detection of the common cystic fibrosis and β-thalassaemia mutations (14). The

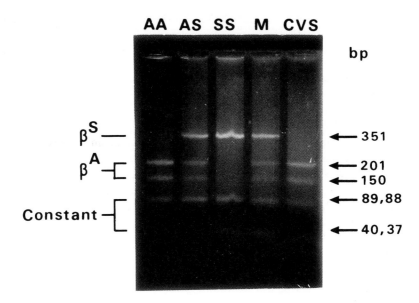

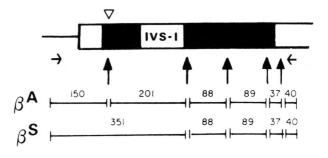

Figure 4 Prenatal diagnosis of sickle cell anaemia by PCR and digestion of the amplified product with *Dde*I. An ethidium bromide stained gel (4%: 3 parts Nuseive agarose, 1 part agarose) shows the digestion products from normal DNA (AA), DNA from a heterozygous individual (AS), a homozygous individual (SS), maternal DNA (M), and CVS DNA. The expected fragment sizes are indicated by the map of the *Dde*I sites in normal DNA (β^A) and DNA with the sickle mutation (β^S).

allele-specific primer has its 3′ terminal nucleotide complementary to either the point mutation (mutant ARMS primer) or the normal sequence at that point (normal ARMS primer), and is designed to amplify DNA only when its 3′ terminal nucleotide forms a perfect template with the target DNA. A second pair of primers is always included in the reaction mixture in order to simultaneously amplify an unrelated DNA sequence as a control to the

ARMS analysis. After amplification the products are simply electrophoresed in an agarose gel and visualized by ethidium bromide staining. A prenatal diagnosis for two β-thalassaemia point mutations using ARMS primers is shown in *Figure 4*. If both parents carry the same mutation then the fetal DNA must be analysed separately with the mutant and normal ARMS primer for that particular mutation.

7.2.6 Chemical cleavage analysis

Point mutations can also be detected by chemical cleavage analysis of hetero-duplexes formed between amplified mutant and normal DNA (15). This approach has been used successfully for carrier and prenatal diagnosis of haemophilia B (16). Amplified normal DNA is end-labelled and annealed to amplified mutant DNA. The heteroduplexes containing a mismatch at the site of the mutation are cleaved by treatment with either osmium tetroxide or piperidine. The end-labelled strand is reduced to a characteristic size for the particular mutation and can be analysed by polyacrylamide gel electrophoresis. However the exact nature of the mutation can only be determined by subsequent direct DNA sequencing.

7.2.7 Denaturing gradient gel electrophoresis

This is an alternative technique for detecting mismatches in amplified hetero-duplex DNA (17). The technique is based on the principle that a double stranded DNA fragment when subjected to electrophoresis through a linearly increasing denaturing gradient will reach a certain point in the gradient where its lowest melting temperature domain melts to create a branched molecule which effectively does not move any further through the gel matrix. This point is determined by the DNA sequence and also the presence of any mismatches in the lowest melting domain of the amplified fragment. To observe the effect of mismatches in the highest melting domain, this part of the DNA sequence has to be converted to the lowest melting domain by the addition of a G + C sequence (GC clamp) at one end of the amplified fragment. Primers and strategies to detect mutations in most parts of the β-globin gene have now been worked out (18).

8. Potential problems

Prenatal diagnosis by DNA analysis has many potential pitfalls, including maternal DNA contamination, contamination of DNA samples with plasmid probes, technical difficulties with the technology, and non-paternity.

8.1 Maternal DNA contamination

Chorionic villus samples are usually contaminated with maternal decidua which must be identified under a low-power dissecting microscope and removed.

Although it has been assumed that a small percentage of contaminating maternal DNA would not significantly contribute to the signal generated by the fetal DNA, several misdiagnoses have been reported as a result of maternal DNA contamination, and widespread adoption of PCR-based techniques with their increased sensitivity causes even greater concern. Contamination can be monitored by the analysis of hypervariable DNA polymorphisms by either Southern blot analysis or DNA amplification. The mother is usually heterozygous for such a polymorphism and often has different sized bands to the father. In such cases any contaminating maternal DNA will show up in the fetal DNA as a weaker third band identical to the second maternal DNA fragment. Such a study in my laboratory on 133 chorionic villus samples sorted by microscopic dissection revealed only one contaminated sample. However the analysis of 28 samples which had been sorted only by eye revealed 10 samples with maternal DNA contamination although the levels were not sufficient to cause a misdiagnosis by Southern blot analysis, the method used at the time.

However this may not be true for PCR analyses. In a PCR analysis, the exponential rate of accumulation of product stops once about one picomole of product has been synthesized. This phenomenon is termed the plateau effect and occurs during the late PCR cycles. A consequence of reaching the plateau effect is that an initially low concentration of other DNA products may continue to amplify preferentially. Thus care should be taken to ensure that the number of cycles are kept to the minimum required to make a diagnosis. It has been reported that levels of maternal DNA up to five per cent of the total do not lead to significant false-positive amplification in the multiplex assay of Duchenne muscular dystrophy deletions as long as the reactions do not approach saturation (11). However PCR analyses for detecting point mutations such as dot-blotting or ARMS may be more sensitive and thus it is most important to amplify for only the minimum number of cycles required.

8.2 Plasmid and target DNA contamination

Contamination of DNA samples or stock solutions with plasmid probes has been reported to have caused at least two misdiagnoses by the Southern blot technique. Plasmid contamination is a problem which is encountered from time to time in most laboratories in which plasmid probes are grown and prepared in the same working area as genomic DNA samples are processed. This problem is identified by the presence of strange bands on the autoradiograph which are often so strong as to obscure weaker-hybridizing genomic DNA fragments. Whenever possible it is recommended that genomic DNA fragments are used as probes instead of whole plasmids. However once the problem arises, tracing the source of the contamination can be difficult and the quickest remedy is to remake every stock solution from its basic ingredients and repeat the analysis.

✳ 130 ng DNA → 1 pmole of 200 bp fragment

The remarkable sensitivity of PCR requires that even greater precautions must be taken when carrying out PCR analyses, because a single contaminating molecule of target sequence DNA from a previous amplification may be amplified leading to a false-positive result. Thus extra precautions should be taken to avoid contamination. The solutions and reaction mixtures for PCR experiments should be made up in a separate area of the laboratory. Amplified product should never be handled in this area. The reaction mixtures must be prepared with a set of pipettes that are never used to handle amplified product. To prevent aerosol contact between the barrel and solution the positive displacement type of pipettes with tips containing disposable plungers are used by some workers instead of the ordinary sort. It is advisable to aliquot reagents and prepare new mixes of buffers for each set of experiments. Solutions and tubes can be irradiated with ultraviolet light before the addition of the target DNA to destroy any other contaminating DNA sequences. When amplifying DNA from very small numbers of cells such as fetal cells in maternal blood or embryonic cells these precautions must be taken to the extreme and it is advisable to set up the reactions in a laminar-flow hood.

8.3 Technical difficulties

One important problem for both Southern blot and PCR analyses is the incomplete (partial) digestion of DNA by restriction endonucleases. This can occur through the use of insufficient enzyme, the wrong reaction buffer or conditions, or the presence of inhibitors which reduce the endonuclease activity. When more than one cleavage site is present in the DNA sequence under amplification or when the fragments analysed by Southern blotting are relatively small, a partial digest will result in an obvious multiple band pattern and the analysis can be repeated. However if there is only one site present in the fragments under study a partial digest may be missed, as in one reported case of a misdiagnosis for β-thalassaemia. DNA samples which give partial digests should be repurified by phenol extraction and ammonium acetate precipitation. The addition of spermidine at a final concentration of 1 to 4 mM in the reation buffer also helps to overcome partial digestion of both DNA samples and amplified fragments. The failure of a DNA sample to amplify while all the control DNA samples work is another occasional problem. A quick remedy is to repeat the amplification using a $10 \times$ diluted aliquot of the DNA sample. If this fails we usually then reprecipitate the DNA with ammonium acetate and dissolve it up in a smaller volume to obtain a purified and more concentrated sample, and try again. Finally various enhancers may be added to the reaction mixture to help overcome difficulties of amplification. These include dimethylsulphoxide (1–10%), glycerol (10–15%), tetramethylammonium chloride (0.05 mM), spermidine (1–4 mM), and formamide (1–10%).

8.4 Non-paternity

The incidence of non-paternity, when a woman's husband or partner is not the biological father of her baby, although not precisely known is significant enough to cause errors in prenatal diagnosis programmes and linkage analysis of large families. In our experience of thalassaemia prenatal diagnosis there have been several diagnoses in which the inheritance of polymorphic markers did not make sense or match the inheritance of β-thalassaemia mutations. In each case non-paternity was demonstrated by genetic fingerprinting with probes for hypervariable regions. Some of these sequences can now be analysed by PCR techniques which makes the routine screening for all prenatal diagnoses a possibility (19).

References

1. MRC working party on the evaluation of chorion villus sampling. (1991). *Lancet,* **337,** 1491.
2. Old, J. M. and Higgs, D. R. (1983). In *Methods in Haematology* (ed. D. J. Weatherall), Vol. 6, pp. 74–102. Churchill Livingstone, Edinburgh.
3. Rebello, M. T., Hackett, G., Smith, J., Loeffler, F. E., Robson, S., MacLachlan, N., Beard, R. W., Rodeck, C. H., Williamson, R., Coleman, D. V., and Williams, C. (1991). *Prenatal Diagnosis,* **11,** 41.
4. Bianchi, D. W., Stewart, J. E., Garber, M. F., Lucotte, G., and Flint, A. F. (1991). *Prenatal Diagnosis,* **11,** 523.
5. Lo, Y.-M. D., Patel, P., Sampietro, M., Gillmer, M. D. G., Fleming, K. A., and Wainscoat, J. S. (1990). *Lancet,* **335,** 1463.
6. Mercier, B., Gaucher, C., Feugas, O., and Mazurier, C. (1990). *Nucleic Acids Res.,* **18,** 5908.
7. Maniatis, T., Sambrook, J., and Fritsch, E. F. (ed.) (1989). *Molecular Cloning. A Laboratory Manual.* Second Edition. Cold Spring Harbor Press, Cold Spring Harbor, NY.
8. Innis, A. M., Gelfand, D. H., Sninsky, J. J., and White, T. J. (ed.) (1990). *PCR Protocols.* Academic Press Inc., San Diego, California.
9. Matthew, C. G. (ed.) (1991). *Protocols in Human Molecular Genetics.* Humana Press, Clifton, New Jersey.
10. Saiki, R. K., Scharf, S., Faloona, F., Mullis, K. B., Horn, G. T., Erlich, H. A., and Arnheim, N. (1985). *Science,* **230,** 1350.
11. Chamberlain, J. S., Gibbs, R. A., Ranier, J. E., Nguyen, P. N., and Caskey, C. T. (1988). *Nucleic Acids Res.,* **16,** 11141.
12. Abbs, S., Yau, S. C., Clark, S., Matthew, C. G., and Bobrow, M. (1991). *J. Med. Genet.,* **28,** 304.
13. Newton, C. R., Graham, A., and Hepstinall, L. E. (1989). *Nucleic Acids Res.,* **17,** 2503.
14. Old, J. M., Varawalla, N. Y., and Weatherall, D. J. (1990). *Lancet,* **336,** 834.
15. Cotton, R. G. H., Rodrigues, N. R., and Campbell, R. D. (1988). *Proc. Natl. Acad. Sci. U.S.A.,* **85,** 4397.

16. Montandon, A. J., Green, P. M., Gianelli, F., and Bentley, D. R. (1989). *Nucleic Acids Res.,* **17,** 3347.
17. Myers, R. M., Maniatis, T., and Lerman, L. S. (1987). In *Methods in Enzymology* (ed. R. Wu), Vol. 155, pp. 501–27. Academic Press, New York.
18. Losekoot, M., Fodde, R., Harteveld, C. L., Van Heeren, H., Giordano, P. C., and Bernini, L. F. (1990). *Br. J. Haematol.,* **76,** 269.
19. Decorte, R., Cuppens, H., Marynen, P., and Cassiman, J.-J. (1990). *DNA Cell Biol.,* **9,** 461.

2

The use of synthetic oligonucleotides as specific hybridization probes in the diagnosis of genetic disorders

SWEE LAY THEIN, ALI EHSANI, and R. BRUCE WALLACE

1. Introduction

This chapter deals with the use of synthetic oligonucleotides as probes in the detection of DNA sequence variations and point mutations. For practical purposes, point mutations can be considered to be single base substitutions, minor insertions, or deletions. To detect these changes in the DNA sequence directly, one exploits the principle that a perfectly matched hybrid formed between the target sequence and an oligonucleotide probe is thermally more stable than one with a single base pair (bp) mismatch. In this respect the design of the allele-specific oligonucleotide (ASO) is critical in that it has to be long enough so that the sequence detected is unique in the human genome and yet short enough so that the hybridization is sensitive to minor differences between probe and template.

The methodology of oligonucleotide hybridization was first developed and used in the detection of the single base mutation in sickle cell disease (1). Subsequently, it has been successfully applied in the diagnosis of the various β-thalassaemia mutations (2, 3), other genetic disorders and malignancies (4), and has become a powerful diagnostic tool. Due to the complexity of human genomic DNA and the inefficiencies in hybridization using short oligonucleotide probes, large amounts (10–15 micrograms) of DNA and highly radioactive ASOs were required. However, it is now possible to achieve a very high degree of enrichment of the target sequence by *in vitro* amplification of genomic DNA using the polymerase chain reaction (PCR) (5). The substantial reduction in complexity of genomic DNA and increased sensitivity has greatly simplified the methodology; only nanogram amounts of sample DNA are required in the PCR, the amplified DNA region of interest is 'dot-blotted' on to a membrane, which can then be hybridized to moderately radioactive ASOs (labelled with ^{32}P or ^{35}S) or ASOs labelled with non-radioactive chemicals. Furthermore, ASOs of shorter lengths allowing increased sensitivity of differentiation between the perfectly matched and single base mismatched hybrids, can be used.

2. Principle of oligonucleotide hybridization

The use of synthetic oligonucleotide as hybridization probes is based on two principles.

(a) That these molecules are capable of forming hydrogen bonds with complementary DNA or RNA sequences giving rise to their specific hybridization behaviour.

(b) That the oligonucleotide–complementary DNA duplex formation is reversible.

The stability of a DNA duplex formation depends on several factors including temperature, ionic strength, base composition, the length of the duplex, the presence of any destabilizing agents, and the presence of any mismatched base pairs. An index of this stability is given by the melting temperature (T_m, temperature at which 50% of the duplexes are dissociated) which is a function of the various factors mentioned. The higher the T_m the more stable is the duplex. In the use of oligonucleotide probes, by comparing the dissociation of several oligonucleotide–DNA complexes as a function of temperature, an empirical formula has been derived (6) for the dissociation temperature (T_d), the temperature at which half of the duplexes are dissociated.

$$T_d \; (^{\circ}C) = 4 \; (G + C) + 2 \; (A + T)$$

where G, C, A, and T indicate the number of the corresponding nucleotides in the oligomer. This relationship is valid only for perfectly matched duplexes between 11 and 20 bases long in 1 M Na^+, and serves as a guide for determining an appropriate hybridization temperature. For hybrids shorter than 20 bp the T_d decreases between 5°C and 10°C for every mismatched base pair. Thus, a temperature 5°C below the T_d is generally used to select for the formation of perfectly matched duplexes, those with one or more mismatched base pairs do not form (7).

Tetramethylammonium chloride (Me_4NCl) which eliminates the dependence of T_d on the G.C content of the probe, can be introduced either in the hybridization or post-hybridization wash (8). Me_4NCl binds selectively to A.T base pairs eliminating the preferential melting of A.T versus G.C base pairs, allowing the stringency of the hybridization to be controlled as a function of probe length only. The method is advantageous for screening a complex library or DNA samples with a pool of oligonucleotide probes.

The time required for hybridization is much shorter for oligonucleotide probes than that for complementary probes, (e.g. those generated by nick translation or hexamer priming). This is due to the much higher molar concentration of the oligonucleotide probe. Unlike complementary probes where probe reannealing in solution can decrease the concentration of probe available for hybridization with the target sequence, oligonucleotide probes are single stranded and thus available in vast excess over that of the target sequences. The hybridization time for oligomers is further minimized as they are short and of low complexity. Thus, for a 15-mer oligonucleotide present at 10 ng/ml, the hybridization is half complete ($t_{1/2}$) in 250 seconds and 99.6% complete in 33 minutes.

Another difference in the hybridization behaviour of the oligonucleotide probes from that of complementary probes is the effect of dextran sulphate on the hybridization rate. The rate of hybridization is increased up to 100-fold for hybridization using complementary probes in dextran sulphate whereas dextran sulphate has a minimal effect on the hybridization rate of short oligonucleotides. This difference is attributed to the ability of complementary probes to form 'networks' between their partially overlapping sequences. The short oligonucleotides being single stranded do not form such networks.

3. Oligonucleotide probe design

In their application for the detection of point mutations, ASO's are used in pairs; one oligonucleotide is completely homologous to the normal sequence, the other to the mutant sequence at a region around the point mutation. Four aspects of the DNA sequence should be considered in the design of oligonucleotide probes: length, G + C content, the presence of non-complementary bases, and self-complementarity.

The length of the oligonucleotide probe determines its hybridization specificity. The longer a sequence, the more likely it is to be unique amongst the collection of sequences the oligonucleotide is used to probe. To be unique the oligonucleotide must contain at least 'N' nucleotides where

$$4^N > 2 \times \text{ no. of bases in the target genome (9).}$$

The haploid human genome contains 3×10^9 bases giving N a value of 17. In addition to specificity, oligonucleotide length determines duplex stability. Sequences shorter than 11 bases have been found to give unacceptable results as probes. However, in deciding the length, we also have to take into consideration the effect of mismatches on the duplex stability to obtain the maximum difference in hybridization between mismatched and perfectly matched duplexes. Previous experience has shown that the optimum length of oligonucleotide for detecting single base mismatches in human genomic DNA is between 19 and 21 bases. However, oligonucleotides of 15–17 bases are adequate for probing amplified DNA sequences.

It has long been established that the higher the GC content of a duplex the greater is the stability of the duplex. Although no detailed thermodynamic study has been done, an empirical relationship of the GC content on duplex stability (given by the parameter T_d) has been derived for oligonucleotide 11–20 bases long in 1 M Na$^+$.

$$T_d \ (^\circ C) = 4 \ (G + C) + 2 \ (A + T)$$

This relationship is useful for estimating the effects of length and G + C content on duplex stability as well as for determining an appropriate hybridization temperature (see earlier discussion).

The presence of a single base pair mismatch will reduce the stability of oilgonucleotide–DNA duplexes. This effect depends on two factors.

(a) Position of the mismatch relative to the end of the duplex. To achieve the maximum destabilizing effect, a mismatch should be central; if this is not possible, it should be at least five nucleotides from either end of a 19 base long oligonucleotide–DNA duplex.

(b) Type of mismatch. For example, the relative stability of G–T versus A–C mismatches has been observed in several applications (10). Therefore, to obtain the maximum destabilizing effect, a G–T mismatch should be avoided whenever possible, and this is achieved by simply synthesizing the complementary sequence producing an A–C mismatch instead. Similarly, it has also been observed (11) that A–G mismatches are more stable than T–C mismatches.

Self-complementarity of oligonucleotides should also be avoided if possible since this may affect hybridization to the complementary DNA and/or the efficiency of labelling.

4. Detection of target DNA sequence

Previously, for maximum sensitivity the labelled ASO's were hybridized to fragmented genomic DNA which has been immobilized *in situ* in a dried agarose gel (12). This approach has now been greatly simplified by the introduction of the PCR (5). The DNA region of interest is enzymatically amplified by the PCR using specific amplification primers and an aliquot of the PCR product is used in the preparation of the filter for hybridization.

Protocol 1. Preparation of filter

1. Subject 1 μg genomic DNA to 30 cycles of PCR amplification. The cycling parameters and buffer conditions should have been chosen such that only a specific fragment is amplified.

2. After completion of PCR, remove the mineral oil. Load 5 μl of the reaction, together with 5 μl of PCR product from a positive and a negative control, and 250 ng of λ *Hind*III or φX174 *Hae*III (whichever is appropriate), in a 1.0% agarose gel. After electrophoresis, stain the gel

with ethidium bromide and examine to see whether amplification was satisfactory.

3. Adjust the amounts of each PCR product, including the negative control, such that it is equivalent to 5 µl of the PCR product from the positive control.

4. For each PCR product, make up to 177 µl with H_2O. To denature, add 10 µl of 500 mM EDTA pH 8.0 and 13 µl of 6 M NaOH to give a final concentration of 2.5 mM EDTA, 0.26 mM NaOH. Stand on ice for 10 min.

5. Meanwhile, soak pre-cut nitrocellulose or nylon membrane in water for 10 min. Place the membrane on the dot-blot apparatus (for example, BRL Dot-Blot manifold) and turn on the vacuum for 1 min.

6. Apply 200 µl of 2 M sodium acetate pH 5.4 to the membrane and turn on the vacuum for 1 min or until all the sodium acetate has been aspirated.

7. Apply the denatured samples on to the membrane and turn on the vacuum again for 1 min.

8. Repeat step 6.

9. Rinse the membrane in 2 × SSC.[a] Blot with Whatman 3MM filter and allow to air-dry.

10. Fix the DNA by baking for 1–2 h at 80°C or by UV illumination.

11. As an alternative to the 'dot-blot', the PCR products could be separated in a 1.0% agarose gel and the DNA transferred on to a membrane by Southern blotting. DNA is then fixed as for step 10.

---

[a] 1 × SSC is standard sodium citrate, 0.15 M NaCl, 0.015 M trisodium citrate, pH 7.0. 20 × SSC stock solution: 175.3 g NaCl and 88.2 g sodium citrate/litre H_2O. Adjust pH to 7.0 with NaOH.

4.1 Preparation of oligonucleotide probe

Oligonucleotides synthesized by the phosphoramidite approach contain a 5'OH which could be phosphorylated with a ^{32}P from [γ-^{32}P]ATP in a kinase reaction using T4 polynucleotide kinase. This method is sufficient for most applications of oligonucleotide probe hybridization. As only one molecule of ^{32}P is incorporated into one molecule of oligonucleotide, the maximum specific activity of the oligonucleotide probe attainable would be that of the [γ-^{32}P]ATP used.

The method of probe purification depends on their eventual use. Since the efficiency of the kinase reaction is not 100%, a proportion (5–10%) of the oligonucleotide remains unlabelled or 'cold'. A clear separation of the labelled (or 'hot') oligonucleotide from the unincorporated [γ-^{32}P]ATP and the 'cold' oligonucleotide can be achieved by thin layer chromatography or polyacrylamide gel electrophoresis in the presence of 7 M urea. The latter was routinely used in our laboratories; in this method the separation depends on the different mobilities of the 5′ phosphorylated and the 5′OH-oligonucleotides. However, for the detection of point mutations in DNA in which the target sequence has been enzymatically amplified by the PCR, the specific activity of the oligoprobe is less critical and it is not necessary to separate the 'hot' oligonucleotide from the 'cold' oligonucleotide. For routine purposes, a clean separation of the labelled oligonucleotides from unincorporated [γ-^{32}P]ATP can be achieved by spun column chromatography using Sephadex G25–50 as the gel filtration medium.

Protocol 2. 5′^{32}P end-labelling

1. Add in the following order:
   - 15 pmol of oligonucleotide[a] (~100 ng for 19-mer)
   - H$_2$O to bring total reaction volume to 10 μl
   - 1 μl of 10 × kinase buffer[b]
   - 1 μl of [γ^{32}P]ATP[c]

   Mix.

2. Add 2 U of T4 polynucleotide kinase. Mix and incubate at 37°C for 30–60 min.

3. Stop the reaction by adding 90 μl of TE buffer[d] to bring the total volume to 100 μl, and leave on ice.

---

[a] Determination of oligonucleotide concentration. Since the base composition of the different oligonucleotides can vary widely, it is necessary to calculate the molar extinction co-efficient (MEC) for each particular sequence in order to determine the accurate concentration. The MEC at 260 nm (pH 8.0) is obtained by summing the contribution of each nucleotide: G12010, A15200, T8400, and C7050.

$$\text{Concentration (mol/litre)} = \frac{\text{OD}_{260}}{\text{MEC}}$$

Oligonucleotides should be made in TE buffer and stored frozen at −70°C.

[b] 10 × kinase buffer is 670 mM Tris–HCl pH 8.0, 100 mM MgCl$_2$, 100 mM dithiothreitol (DTT).

[c] [γ^{32}P]ATP that we use is from Amersham (PB15068 ~3000 Ci/mmol).

[d] TE buffer is 10 mM Tris–HCl pH 8.0, 1 mM EDTA.

Protocol 3. Separation of oligonucleotide probe by Sephadex G25–50 spun column chromatography

1. Remove the plunger from a 1 ml syringe and plug the bottom with glass wool. Place the syringe in a 15 ml Falcon tube so that the finger grips of the syringe hang from the rim of the tube.

2. Fill the syringe with pre-swollen Sephadex G25–50 previously equilibrated with TE buffer. Spin for 10–30 sec at 1000 r.p.m. in a bench-top centrifuge,

(e.g. Sorvall RT6000). The Sephadex will pack down. Discard eluate. Add more Sephadex and spin again. Continue until the bed volume is ~1 ml.

3. Position a 'headless' 1.5 ml Eppendorf in the Falcon tube. Load 100 μl TE and spin at 1000 r.p.m. for 4 min.

4. Repeat as necessary until the column is equilibrated, i.e. 100 μl is recovered in Eppendorf.

5. Load the kinase reaction (which should have been adjusted to 100 μl with TE) on to the column and spin under identical conditions. 100 μl of the flow through containing the labelled ASO should be recovered.

   (*Caution*: it is important not to vary the centrifugation conditions as any change can lead to incomplete recovery of the sample.)

6. Add 100 μl TE to the recovered probe to bring the total volume to 200 μl. Take 2 μl, add to 5 ml scintillation fluid, and count.

4.2 Oligonucleotide probe hybridization

Generally, all hybridization and washing manipulations of oligonucleotide probes are performed in 1 M Na^+ conditions so that the stringency can be altered by the temperature of hybridization, and the temperature and time of the post-hybridization wash. These conditions vary with the length and sequence complexity of each ASO and should have been worked out for each set of oligoprobes using positive and negative DNA controls. It should be pointed out that some background hybridization will be present so that the most important parameter in ASO probe hybridization is the 'signal-to-noise' ratio. In the procedure outlined here, the initial wash is followed by a Me_4NCl wash to improve the selectivity of hybridization.

Protocol 4. Hybridization and washing

1. Pre-wet the membrane in 1 × SSC and pre-hybridize in heat-sealed plastic bags for a minimum of 30 min in 5 × SSPE[a], 0.1% SDS, and 100 μg/ml tRNA at the appropriate temperature (usually 5–10°C below T_d of the ASO). Approximately 10 ml of solution per 100 cm^2 membrane should be used. Squeeze as much air as possible from the bag before sealing.

2. Remove the bag from the water-bath. Open the bag by cutting a corner and add the oligonucleotide probe (~2 × 10^6 c.p.m./ml) directly to the mix. Squeeze as much air as possible from the bag and reseal.

3. Hybridize for a minimum of 2 h at the temperature of pre-hybridization.

4. Remove the bags from the water-bath, dry the outside, and cut off one corner. Pour the hybridization buffer into a beaker for disposal. Cut the bag along the length of three sides and remove the membrane. Immediately submerge the membrane in 6 × SSC.

Protocol 4. *Continued*

5. Wash the membrane in 6 × SSC with gentle agitation for 20 min at room temperature. Repeat.

6. Rinse the membrane in T-MAC^b wash solution. Transfer the membrane to a lidded box and wash in T-MAC solution for 20 min. Repeat.

   (*Caution*: the actual temperature of this final wash should be monitored carefully. For a 19-mer ASO, a wash temperature of 54–56°C would be suitable.)

7. Rinse the washed membrane in 2 × SSC and wrap the damp membrane in cling-film. Autoradiograph with Kodak XAR film between two intensifying screens at −70°C for 1–16 h.

   (*Caution*: do not allow membrane to dry out at any stage as additional washes may be required.)

8. The membrane can be re-hybridized with a different probe after elution of the previous probe. To remove previous probe incubate the membrane in 5 mM Tris–HCl pH 8.0, and 0.1% SDS at 75°C for ~15 min. Rinse with 2 × SSC at room temperature. Autoradiograph to check that all the probe has been removed. The membrane may now be re-hybridized with a different probe.

^{a} 1 × SSPE is 180 mM NaCl, 10 mM NaH_2PO_4, 1 mM EDTA (pH 8.0).
^{b} T-MAC wash: 3 M tetramethylammonium chloride [$(CH_3)_4NCl$] (Aldrich TI, 952–6), 50 mM Tris–HCl pH 8.9, 2 mM EDTA, 0.1% SDS. The T-MAC stock solution is made up by dissolving the Me_4Cl in H_2O to a concentration of 4 to 4.5 M. The exact concentration may be calculated by measuring the refraction index (R.I.) of this solution at 20°C using the following formula:

$$\text{Conc. (M)} = \frac{\text{R.I.} - 1.331}{0.018}$$

Caution: Me_4Cl is extremely hygroscopic.

4.3 An alternative hybridization—competition hybridization

One potential disadvantage of direct hybridization with ASOs for the discrimination of allelic genes or the transcripts of different alleles, is the necessity to determine the exact conditions under which discrimination is possible. In the case of G–T or G–U mismatches formed between the ASO probes and β-globin mRNA, discrimination was found to be difficult to achieve (13). One solution to this problem is to use a competition hybridization approach (13, 14).

Competition hybridization involves hybridization with a labelled ASO in the presence of unlabelled competitor oligonucleotide(s). The competitor oligonucleotide usually differs from the labelled probe by a single nucleotide and is added in at least a ten-fold molar excess. If the sequences of two alleles are known, the probe to allele one is labelled and the unlabelled probe to allele two is added to suppress any hybridization of the labelled probe to its non-complementary target sequence (14). Including competitors in the hybridization mixture eliminates the need of establishing stringent hybridization and wash conditions. Hybridization and washes can be done under standard conditions for most probes.

A further advantage of competition hybridization is that the oligonucleotide probes can be longer allowing a higher specificity. Because discrimination is achieved by competition instead of by the kinetics of dissociation, even one mismatch in 23 nucleotides is sufficient difference to permit the design of discriminating probes (14).

Protocol 5. Competition hybridization

1. Amplify target sequence by PCR as for *Protocol 1*.

2. Denature 1 µl of PCR reaction by diluting with 400 µl 10 mM NaOH, 1 mM EDTA. Filter through Genetran membrane in a dot-blot apparatus using vacuum. Wash with an equal volume of denaturing solution. Rinse membrane in 2 × SSC and blot dry.

3. Cross-link with UV light, (e.g. use Stratalinker Model 1800 at 1200 microjoules for 1 min).

4. Hybridize the membrane for a minimum of 1 h at the appropriate temperature (10°C below calculated T_d) with 10^6 c.p.m./ml of 5′ end-labelled oligonucleotide in 5 × SSPE, 0.1% SDS, 10 µg/ml tRNA. The labelled probe is specific for one allele. Add unlabelled oligonucleotide specific for the same region of the other allele(s) at a concentration of 20 pmol/ml.

5. Remove the blot and wash for 30 min in 6 × SSC at room temperature. Repeat.

6. Wash for 1–2 min (probes < 20 nt) or 5 min (probes > 20 nt) in 6 × SSC at the hybridization temperature.

7. Wash again for 15 min in 6 × SSC at room temperature.

8. Dry the blot between two sheets of Whatman 3MM paper and wrap the blot in cling-film and autoradiograph with Kodak XAR-5 film between two intensifying screens (Cronex Lightning Plus) at −70°C for 1 h.

9. Repeat steps 5 and 6 as necessary to obtain specificity of hybridization. It is therefore necessary to always include negative and positive hybridization controls.

---

5. Practical applications

Examples of ASOs in the detection of point mutations in the human genome are numerous (2–4). In this section we illustrate the application of competition hybridization in the analysis of polymorphisms in the *c-erb-B2* transmembrane region.

5.1 Polymorphism in the *c-erb-B2* transmembrane region

Competition hybridization can be used to analyse for the presence of four possible alleles of the *c-erb-B2* oncogene. *Figure 1* shows the sequence of the transmembrane region of the gene. This region can be amplified by PCR using the upper and lower strand primers shown by the boxes. The four allele specific oligonucleotide probes EB TT, EB CT, EB CC, and EB TC are used in competition hybridization to determine which alleles are present in the genome of the individual. To date, the allele to be detected by the EB TC probe has never been detected (> 500 chromosomes analysed).

5.2 Specificity of competition hybridization

Four different sequences representing possible alleles of the transmembrane region of the *erb-B2* oncogene were synthesized by PCR mediated *in vitro* mutagenesis. Template (5′CGCAGAGATGATGGACGTCATGACTTTT-ATGCCCAGCC) (0.1 fmol) was amplified in a reaction containing 50 mM KCl, 10 mM Tris–HCl pH 8.3, 1.5 mM MgCl$_2$, 0.1% gelatin, 200 µM dNTPs, 10 pmol BGP1 (5′GGGCTGGGCATAAAAGTCA), 10 pmol of *c-erb-B2* primer (EB TT, EB CT, EB CC, or EB TC, see *Figure 2*), and 1.5 U Amplitaq (Cetus). Reactions were carried out for 20 cycles at an annealing temperature of 37°C for 30 seconds, a polymerization temperature of 72°C for 30 seconds, and a denaturation temperature of 94°C for one minute. PCR

5′ GA|GAGCCAGCCCTC|TGACGTCC|ATCATCTCTGCCG|GTGGTTGGCATTCTGCTGGTCGTGGTCTTGGGGGTGGTCTTTGGGATCCTCATCAAGCGACGG
CTCTCGGTCGGGAG|ACTGCAGG|TAGTAGAGACGC|CACCAAGGGTAAGACGACCAGCACCAGAACCCCCACCA|GAAACCCTAGGAGTAGTTCG|CTGCC 5′

ASO

ASO PROBES:

EB TT   CGCAGAGA*T*GA*T*GGACGTCA

EB CT   CGCAGAGA*C*GA*T*GGACGTCA

EB CC   CGCAGAGA*C*GA*C*GGACGTCA

EB TC   CGCAGAGA*T*GA*C*GGACGTCA

Figure 1 Analysis of polymorphism in the *c-erb-B2* transmembrane region.

PROBE:

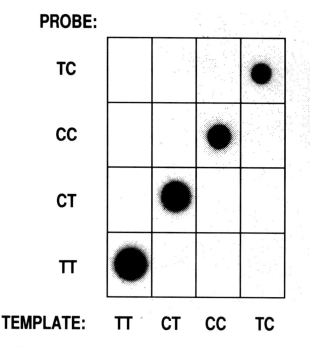

TEMPLATE: TT   CT   CC   TC

Figure 2 Specificity of competition hybridization.

products (1 μl per well) were mixed with 400 μl of 10 mM NaOH, 1 mM EDTA and filtered on to a Genetran membrane under vacuum in a dot-blot apparatus. After filtration, the samples were washed with 400 μl of the denaturing solution. The filter was then cross-linked by UV light in a Strata-linker Model 1800 at 1200 microjoules for one minute. Four identical filters each containing the four possible amplified templates were prepared in this way. Each filter containing the four templates was hybridized with one of the ASO probes, ED TT, EB CT, EB CC, or EB TC by competition hybridization. The kinased labelled probe (10^6 c.p.m./ml) was mixed with hybridization solution (see *Protocol 5*) in the presence of the other three unlabelled probes (20 pmol/ml). Hybridization was at 59°C for four hours. The filters were washed three times in 6 × SSC at room temperature with agitation and in T-MAC wash solution at 59°C for 35 minutes. The filters were then exposed to X-ray film. The signal-to-noise ratio observed for the hybridization is >100:1.

6. Notes

(a) Background hybridization can be a problem. A 'spotty' type background is usually due to probe contamination with unincorporated [γ^{32}P]ATP. A 'blotchy' background is usually due to poor washing. A prolonged wash in 2 × SSC at room temperature following the final wash helps.

(b) Since the efficiency of DNA amplification varies from sample to sample, interpretation of dot-blot hybridization of amplified DNAs can be problematical unless comparable amounts of target DNA are applied in the dot-blots. This is best done by running small aliquots of the PCR products in an ethidium bromide stained minigel.

(c) To obtain probes of comparable specific activity or high specific activity, the 'hot' oligonucleotide should be separated from the 'cold' oligonucleotide in 7 M urea polyacrylamide gel electrophoresis (12).

(d) Oligonucleotide probes with extremely high specific activities (five to ten times those achieved by the kinase reaction) can be obtained by 3' prime extension. This method used the Klenow fragment of *E. coli* DNA polymerase I to incorporate $[\alpha^{32}P]dNTPs$ in a primer/template complex. The newly synthesized labelled strands must be separated by polyacrylamide gel electrophoresis under denaturing conditions (12).

(e) Enrichment of the target DNA sequences by PCR over the rest of the genome has increased the sensitivity of detection in dot-blot hybridization allowing the use of ^{35}S-labelled and non-radioactively labelled probes. For example, the use of probes covalently labelled with horseradish peroxidase (15) allows detection with a colorimetric assay. ^{35}S-labelled probes (16) can be used for up to three months and the non-radioactively labelled probes have a shelf-life of two years.

(f) A modification of the approach outlined in this chapter is the 'reverse' dot-blot in which the ASOs are immobilized on to a detection strip of membrane and the amplified DNA is hybridized to the membrane (17). The approach is particularly valuable where the potential number of mutations or sequence variation is large; for example, in the detection of a particular β-thalassaemia mutation or in HLA typing. Since the dot-blot approach allows the analysis of a large number of individuals with a single probe and the reverse dot-blot approach allows the analysis of a large number of loci of a single individual, choice of approach will very much depend on the ratio of individuals to loci in your study.

Acknowledgement

We thank Liz Rose for excellent help in preparation of the manuscript.

References

1. Conner, B. J., Reyes, A. A., Morin, C., Itakura, K., Teplitz, R. L., and Wallace, R. B. (1983). *Proc. Natl Acad. Sci. U.S.A.*, **80**, 278.

2. Orkin, S. H., Markham, A. F., and Kazazian, H. H. J. (1983). *J. Clin. Invest.*, **71**, 775.
3. Thein, S. L., Wainscoat, J. S., Old, J. M., Sampietro, M., Fiorelli, G., Wallace, R. B., and Weatherall, D. J. (1985). *Lancet,* **ii**, 345.
4. Verlaan-de Vries, M., Bogaard, M. E., Van den Elst, H., van Boom, J. H., van der Eb, A. J., and Bos, J. L. (1986). *Gene, 50*, 313.
5. Saiki, R. K., Scharf, S., Faloona, F., Mullis, K. B., Horn, G. T., Erlich, H. A., and Arnheim, N. (1985). *Science,* **230**, 1350.
6. Suggs, S. V., Hirose, T., Miyake, T., Kawashima, E. H., Johnson, M. J., Itakura, K., and Wallace, R. B. (1981). In *Developmental Biology using Purified Genes. ICN-UCLA Symposia on Molecular and Cellular Biology*, (ed. D. D. Brown and D. F. Fox), pp. 683. Academic Press, New York.
7. Wallace, R. B., Shaffer, J., Murphy, R. F., Bonner, J., Hirose, T., and Itakura, K. (1979). *Nucleic Acids Res.*, **6**, 3543.
8. Wood, W. I., Gitschier, J., Lasky, L. A., and Lawn, R. M. (1985). *Proc. Natl Acad. Sci. U.S.A.,* **82**, 1585.
9. Thomas, C. A. J. (1966). In *Progress in Nucleic Acid Research and Molecular Biology* (ed. J. N. Davidson and W. E. Cohn), pp. 315. Academic Press, New York.
10. Wallace, R. B., Johnson, M. J., Hirose, T., Miyake, T., Kawashima, E. H., and Itakura, K. (1981). *Nucleic Acids Res.*, **9**, 879.
11. Ikuta, S., Takagi, K., Wallace, R. B., and Itakura, K. (1986). *Nucleic Acids Res.,* **15**, 797.
12. Thein, S. L. and Wallace, R. B. (1986). In *Human Genetic Diseases: A Practical Approach* (ed. K. E. Davies), pp. 33–50. IRL Press, Oxford.
13. Nozari, G., Rahbar, S., and Wallace, R. B. (1986). *Gene,* **43**, 23.
14. Wu, D. Y., Nozari, G., Schold, M., Conner, B. J., and Wallace, R. B. (1989). *DNA,* **8**, 135.
15. Saiki, R. K., Chang, C.-A., Levenson, C. H., Warren, T. C., Boehm, C. D., Kazazian, H. H. J., and Erlich, H. A. (1988). *N. Eng. J. Med.,* **319**, 537.
16. Cai, S.-P., Zhang, J.-Z., Huang, D.-H., Wang, Z.-X., and Kan, Y. W. (1988). *Blood,* **71**, 1357.
17. Saiki, R. K., Walsh, P. S., Levenson, C., and Erlich, H. A. (1989). *Proc. Natl Acad. Sci. U.S.A.,* **86**, 6230.

3

Pulsed-field gel electrophoresis in the analysis of genomic DNA and YAC clones

J. T. DEN DUNNEN, P. M. GROOTSCHOLTEN,
and G. J. B. VAN OMMEN

1. Introduction

Since the first appearance of this book in 1986 pulsed-field gel electrophoresis (PFGE) has evolved from a newly developed technique (1) to a fundamental research tool indispensable to a modern human genetics laboratory. Initially, the technique was mainly used to separate very large DNA molecules, but more recently its potential for separating smaller DNA molecules and even its application in DNA sequencing (2) became evident. Applications of PFGE include physical mapping (3–6), size selection (7–9), the study of genetic rearrangements causing cancer and genetic disease (10–12) or underlying immunological variation (13), yeast artificial chromosome cloning (YAC) technology (7, 8, 14, 15), the separation of DNA molecules with specific structures, and applications in human and veterinary microbiology and parasitology (16, 17). To cope with the fast developments, especially of the technology, we had to completely revise the chapter published in the first edition (18).

In recent years, several groups have studied the theoretical basis of electrophoresis through a directionally alternating electric field. Models have been designed and validated both in practice and by computer simulation (19–22). Some of the computer programs are available upon request from their authors. Although the understanding of migration of large molecules through matrices in alternating fields has improved, the value of theoretical studies for the practical design of equipment is still limited. Most equipment commercially available nowadays has been designed largely based on empirical trial and error. This has however yielded a variety of well-functioning systems, which can be readily applied to most purposes. The selection of equipment depends mostly on individual preferences and budgetary (and occasionally laboratory space) considerations.

In this chapter we briefly review the various technologies available, followed by the presentation and in-depth discussion of practical and theoretical aspects of each subsequent step in the long-range analysis of chromosomal DNA and YACs. In the detailed protocols much attention is devoted to the identification and prevention of potential pitfalls. It should be noted that, while we mainly focus on the use of a CHEF-type of system, in fact all other technologies can be substituted with minor modifications related to possibilities and limitations of the system configuration.

2. Available PFGE systems

The first electrophoretic technique that allowed the resolution of DNA molecules above 50 kb was developed by Schwarz and Cantor (1) and was called 'pulsed-field gradient gel electrophoresis'. Nowadays, many systems are available all with variations on the basic principle of a continuous reorientation of the DNA molecules caused by a recurrent change in electric field direction. Different systems vary considerably in gel-box construction, gel dimensions, electrode configuration, and application of electric fields (pulse algorithms). The term pulsed field gel electrophoresis (PFGE) is now used as a generic acronym to indicate any technique which resolves DNA by discontinuous reorientation. The most frequently used systems, often commercially available, are listed below.

i. Field inversions gel electrophoresis
The FIGE system (23) uses normal submarine gel-boxes with a possibility of buffer cooling (*Figure 1C*). Reorientation of the DNA molecules is achieved by reversing the field polarity (reorientation angle 180°) in either alternating switching intervals with roughly a 3:1 ratio for forward to reverse fields, or by the application of higher forward than backward field strengths, (e.g. programmable Power Inverter, MJ Research, USA; Gene-Tic™, Biocent, Netherlands).

ii. Contour-clamped homogeneous electric field electrophoresis
In CHEF electrophoresis (24) the electric field is distributed along the contour of a hexagonal array of electrodes (*Figure 1D*). The opposing sides of the hexagon are activated alternately in a 120° angle. CHEF electrophoresis represents the most popular PFGE system and is available from several companies (Pulsaphor™, Pharmacia, Sweden; CHEF-DR™, Bio-Rad USA; Gene-Tic™, Biocent, Netherlands).

iii. Transverse alternating field electrophoresis
TAFE (25) is unique in the way that it uses a vertically placed gel, perpendicular to the electric field (*Figure 1B*, Geneline™, Beckman, USA). A variant of this system, called ST/RIDE (26) has a fixed electrical field applied to a vertical gel and a simultaneously applied perpendicular field which alternates polarity in a cyclic fashion (ST/RIDE™, Stratagene, USA).

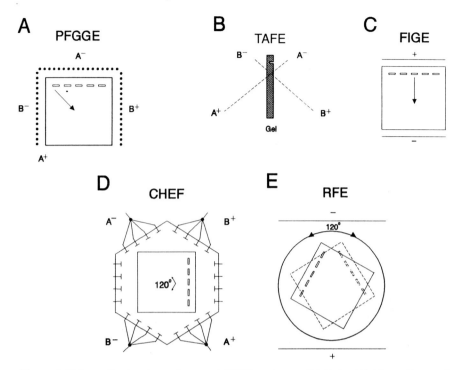

Figure 1 Schematic drawing of several PFGE systems. (A) Pulsed-field gradient gel electrophoresis; the first system described (1). (B) Transverse alternating field electrophoresis; the system using vertical gels. (C) Field inversion gel electrophoresis; the cheapest system able to operate in standard gel-boxes. (D) Contour-clamped homogeneous electric field electrophoresis; the most widely applied system. (E) Rotating field electrophoresis; shown is the RGE variant (see Section 2). The electric field alternates between positions A^-/A^+ and B^-/B^+.

iv. Rotating field electrophoresis

Two types of RFE are available. Rotating gel electrophoresis (RGE) also called the 'Waltzer'-type (27), uses a normal gel-box in which the gel lies on a turntable that moves back and forward under a selectable angle (normally 120°) between two rest positions (*Figure 1E*). In rotating electrode electrophoresis (REE) (Rotaphor™, Biometra, Germany) the gel is stationary but two long wire electrodes turn around the gel (28).

v. Future developments

Recent improvements tend to result in the design of gel-boxes in which a large set of point-electrodes are arranged in a square array. The voltage on every individual electrode can be regulated independently by a computer. In such a set-up, each of the earlier mentioned systems, except the TAFE system, can be simulated by the software which controls the electrodes.

All systems are driven by a programmable controller which provides a recurrent inversion of output polarity and which allows the setting of a variable number of parameters defining the separation obtained. The flexibility of the software driving the electrophoresis is an essential component of the complete system. Not all parameters can be changed in every system.

3. Preparation of DNA samples

3.1 Isolation of DNA in agarose

Essentially, two methods have been described to prepare agarose embedded DNA of sufficient double stranded length. We use the method first described by Schwartz and Cantor (1, 29) of including cells in agarose blocks (plugs), using the mould shown in *Figure 2*. The plugs are easy to manipulate, no centrifugations are required, no 'void volume' exists in buffer changes, and no increased quantities of enzymes are required. Alternatively, the cells are included in agarose beads by making a suspension of the still molten agarose in liquid paraffin and washing the mixture with buffer when the agarose has solidified (30). Centrifugation makes the beads settle and further treatments are essentially identical to those applied for the plugs. We have consistently observed 30–50% lower yields with the beads, probably due to loss of smaller particles and cells or DNA at the surface of the beads. The beads would need shorter equilibration times of buffers and enzymes. However, the applied incubation times render this an insignificant advantage.

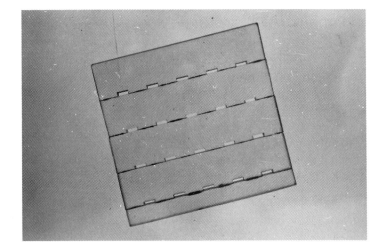

Figure 2 Perspex block mould, made of 10 mm thick Perspex strips in which 6 mm wide and 1.5 mm deep slits were made. Subsequently, several strips were glued together to form the block slots.

In principle it is feasible to prepare high molecular weight DNA in solution, yielding DNA of 100–400 kb. This may be practical for preparative purposes, for example to religate large circles in the preparation of DNA 'jumping libraries' (31) or in the preparation of yeast artificial chromosomes (7). However, this method is very laborious since extreme care needs to be taken to avoid DNA shearing (cut-off pipette tips, gentle mixing, overnight dissolving of samples). Furthermore, when enzymes are used which cut very infrequently, like *Not*I, where most of the digested DNA remains above 500 kb, there is no alternative to using agarose-embedded DNA.

Protocol 1. Preparation of agarose blocks containing human DNA

White blood cells were obtained by isotonic lysis of whole blood (32). Tissue culture cells were collected using standard procedures.

1. Wash the cells once in SE (75 mM NaCl, 25 mM Na EDTA, pH 7.4) and resuspend in SE at 20×10^6 cells/ml, at room temperature.

2. Melt 1% LGT agarose (InCert-agarose, FMC) in SE, cool to about 50°C, and mix in a 1:1 ratio with the cell suspension.

3. Immediately dispense the mixture into slots of $10 \times 6 \times 1.5$ mm (100 μl volume), made through a Perspex mould (see *Figure 2*) and covered on one side with tape.

4. Put the mould on ice for 5–10 min.

5. Remove the tape and blow the solidified blocks gently out of the slots using a Pasteur pipette balloon.

6. Collect the blocks in about five volumes of SarE (1% Sarcosyl, 0.5 M EDTA, pH 9.5).

7. Add proteinase K to 0.5 mg/ml, followed by 30 min incubation at room temperature, and overnight incubation at 55°C. (Proteinase K can be replaced with pronase (1 h pre-incubated) at the same concentration, followed by overnight incubation at room temperature, without noticeable effect on the results.)

8. Rinse the blocks several times with distilled water, and wash at least three times for 2 h and once overnight in 10–20 volumes of TE 0.5 (10 mM Tris, 0.5 mM EDTA, pH 7.4), by gentle rotation.

9. Store the blocks at 4°C in 50 mM EDTA (pH 8.0).

0.1 mM phenylmethylsulphonylfluoride (PMSF) (Sigma) can be added to the first one or two washes to block protease activity (17). A smear throughout the lanes after electrophoresis indicates poor cell lysis or DNA degradation. Remaining DNA degrading activities can be eliminated by a second pronase treatment. In 'emergency cases', (e.g. rare valuable samples) degraded DNA

Protocol 1. *Continued*

can be removed from a plug by a short 'pre'-electrophoresis before further handling of the sample. Before a first digestion, a control incubation should be done without the addition of enzyme. This step should show negligible DNA degradation in the size range under study (< 2 Mb).

3.2 Preparation of marker DNAs

Several markers, most of which are commercially available, are suitable to run in parallel with the DNA samples to be analysed (see *Table 1*). Standard size markers, including lambda digests with *Kpn*I, *Nae*I, *Nru*I, and *Xho*I, can be used in the size range of 10–40 kb. Oligomers of φX174 (available from Pharmacia) have a 5.4 kb spacing and serve as ideal markers for the next level of 25–250 kb. However, for PFGE the most frequently used markers are oligomers of bacteriophage lambda DNA and the chromosomes of the yeast *Saccharomyces cerevisiae*. Lambda markers give a ladder with a 48.5 kb spacing and are optimal in the range of 50–800 kb. *S. cerevisiae* has chromosomes ranging in size from 220–2500 kb, however the largest chromosome can not be used as a reference marker since its size is, even within a given strain, not stable. Furthermore, depending on the *S. cerevisiae* strain used,

Table 1. Sizes (in kb) of PFGE marker molecules

Lambda	S. cerevisiae	K. lactis	C. albicans	S. pombe
0.776	>2.3	>2.6	>3.0	5.7
0.728	1.58	2.2	>2.7	4.6
0.679	1.12	1.8	2.3	3.5
0.631	0.94	1.3	1.9	
0.582	0.83	1.05	1.75	
0.534	0.78		1.7	
0.485	0.74		1.25	
0.437	0.69		1.12	
0.388	0.60 [a]		1.00	
0.340	0.44			
0.291	0.35			
0.243	0.28			
0.194	0.22			
0.146				
0.097				
0.049				

Yeast chromosome sizes are calibrated with bacteriophage lambda CI857Sam7 (genome size 48.5 kb). Yeast strains used are *Saccharomyces cerevisiae* AB1380 (14), *Kluyveromyces lactis* var. lactis CBS2859 (50), *Candida albicans* CBS562 (50), *Schizosaccharomyces pombe* CBS356 (50).
[a] Doublet band.

individual chromosomes vary considerably in length. The chromosomes of *Kluyveromyces lactis* or *Candida albicans* provide useful markers in the size range of 1.0–3.0 Mb. The chromosomes of *Schizosaccharomyces pombe*, 3.5–5.7 Mb, can be used to estimate the sizes of yet bigger DNA molecules.

Protocol 2. Preparation of bacteriophage lambda oligomer plugs

1. Dilute a solution of lambda phage particles to a DNA concentration of 10–20 μg/ml in SE. Mix gently 1:1 with 1% LGT agarose in SE at 45°C, dispense in slots, and allow to solidify, following the procedures mentioned above.
2. Collect the blocks in about five volumes of SarE (1% Sarcosyl, 0.5 M EDTA, pH 9.5).
3. Add pronase to 0.5 mg/ml, followed by 2 h incubation at 37°C, and overnight incubation at room temperature.
4. Wash the blocks and store in 50 mM EDTA (pH 8.0) as described above.

Annealing of the lambda sticky ends depends mainly on DNA concentration in the plugs and temperature during preparation. When the ladders do not reach the desired size range, their length can be increased by a $MgCl_2$ incubation. Equilibrate the plugs to 10 mM $MgCl_2$ and incubate 15 minutes at 42°C. Wash extensively in TE and store in 50 mM EDTA (pH 8.0). Good oligomeric ladders can also be obtained with commercial DNA preparations. For this purpose, pure lambda DNA must be included in agarose blocks at a concentration of 20 μg/ml in SE, followed by annealing for three to five hours at 37°C. Freeze-dried and frozen DNA batches perform poorly, while preparations which have only been stored refrigerated produce similar results as never frozen 'home-made' preparations. Lambda DNA preparations which have been stored refrigerated for extended periods (over one to two weeks) in concentrations over 0.1 mg/ml were found to contain mainly oligomers of 200–600 kb. The handling of these solutions takes great care to avoid shear, e.g. gentle pipetting and swirling in wide test-tubes for mixing.

Protocol 3. Preparation of yeast DNA plugs

In general, 10 ml yeast culture in YPD (1% yeast extract, 2% pepton, 2% dextrose) is used to make 40–80 plugs.

1. Grow the yeast overnight in YPD to early stationary phase, dilute 1:10, and grow to a density of 1.0 OD_{600}.
2. Collect the cells by centrifugation (5 min, 300 g), resuspend in four volumes SE, mix 1:1 with 1% LGT agarose in SE at 50°C, containing 20 mM dithiothreitol (DTT), 30 μg/ml Zymolyase 20 000 (Seikagaku Kogyo Co. Ltd., Japan).

Protocol 3. *Continued*

3. Pour the mixture into the block mould, chill, transfer it to an equal volume of SE–DTT–Zymolyase, and incubate 1.5 h at 37°C.

4. Further process the blocks as described for blocks containing human DNA (see *Protocol 1*, step **6**).

The protocol gives good results for most yeast strains, although some, (e.g. *S. pombe*) may need more stringent conditions to give good cell lysis. In these cases (step **4**) rinse in 0.5% SDS, 0.5 M EDTA (pH 8.0) and incubate overnight at 50°C in two volumes 0.5% SDS, 0.5 M EDTA (pH 8.0) containing 1.0 mg/ml proteinase K.

3.3 Restriction digestion of agarose-embedded DNA

Digestion of agarose-embedded DNA requires conditions similar to soluble digestions. However, PFGE analysis requires the use of specific, infrequently cutting restriction endonucleases (rare-cutters, *Table 2*), modifications of

Table 2. Rare cutting restriction enzymes

Enzyme	Sequence	Enzyme	Sequence
*Asc*I	GG/CGCGCC	*Pvu*I	CGAT/CG
*Asu*II	TT/CGAA	*Rsr*II	CG/GwGCC
*Bss*HII	G/CGCGC	*Sac*II	CCGC/GG
*Cla*I	AT/CGAT	*Sal*I	G/TCGAC
*Eag*I	C/GGCCG	*Sfi*I	GGCC(n)$_4$/nGGCC
*Kpn*I	GGTAC/C	*Sma*I	CCC/GGG
*Ksp*I	CCGC/GG	*Sna*BI	TAC/GTA
*Mlu*I	A/CGCGT	*Spe*I	A/CTAGT
*Nae*I	GCC/GGC	*Spl*I	C/GTACG
*Nar*I	GG/CGCC	*Spo*I	TCG/CGA
*Not*I	GC/GGCCGC	*Tth*111I	GACn/nnGCT
*Nru*I	TCG/CGA	*Xba*I	T/CTAGA
*Pac*I	TTAAT/TAA	*Xho*I	C/TCGAG

The most frequently used rare cutting restriction enzymes (rare-cutters) with their recognition sequences (5'→3') and cutting position (/).

digestion protocols, altered techniques to load the DNA samples, and modification of the techniques to blot and hybridize the DNA.

Protocol 4. Restriction digestion of DNA in plugs

Digestions of DNA in plugs are essentially performed according to the manufacturer's recommendations, generally at two to three-fold enzyme excess.

Half or one-third of a plug is required per lane on a gel ($\pm 5 \times 10^5$ cells/lane). The most frequently used rare-cutters are given in *Table 2*.

1. Cut the plugs (approximate volume 100 μl) in half before use.

2. Rinse the plugs, which were stored in EDTA, once with sterile water, and wash at least three times for 2 h in 10–20 volumes 0.5 TE under gentle rotation to remove all EDTA.

3. Equilibrate each 50 μl plug for 2 h at room temperature or overnight at 4 °C with 1 ml of the appropriate digestion buffer.

4. Replace the wash by 50 μl fresh buffer, to which 4 mM spermidine, 2 mM DTT, and 0.2 mg/ml bovine serum albumin (BSA) (Boehringer, MB grade) have been added.

5. Carry out digestions for 6 h to overnight at the specified temperature, using 10–20 units of enzyme per 50 μl plug. The enzyme is usually added in two equal portions, at the beginning and half-way through the digestion time. Control incubations without enzyme will show if the methods used produce negligible degradation of DNA in the size range under study (up to 2 Mb).

6. After digestion either layer the plugs directly or store until use in 5 mM EDTA at 4 °C. For double digestions, repeat steps 1 to 4 for the second enzyme. For the second digestion step 1 may be preceded by a proteinase K treatment, but should not be necessary. Partial digestions are preceded by an experiment to determine the amount of enzyme required to obtain DNA of the desired length. Step 4 is performed on ice with all solutions cooled to 4 °C. The added fresh buffer contains the required amount of restriction enzyme and the sample is incubated overnight at 4 °C to achieve a homogeneous enzyme concentration by diffusion. Subsequently, the sample is incubated for 30 minutes at the digestion temperature. Digestion is stopped by storage on ice and addition of EDTA (pH 8.0) to 50 mM. For digestions with more frequently cutting restriction endonucleases (like *Eco*RI, *Hind*III) modify the protocol after step 3 to: carefully remove all equilibration buffer and incubate 10 minutes at 65 °C to melt the plug. Incubate 15 minutes at 37 °C, directly add DTT, spermidine, BSA, and restriction enzyme, and incubate at the desired temperature. Layer the melted plug directly on to the gel.

4. Electrophoresis

4.1 The PFGE system

Our PFGE system (see *Figure 3*) uses the Gene-tic power supply (Biocent, P. O. Box 280, 2160 AG LISSE, The Netherlands) to drive the electrophoresis. This power supply (*Figure 3B*) is capable of driving four independently

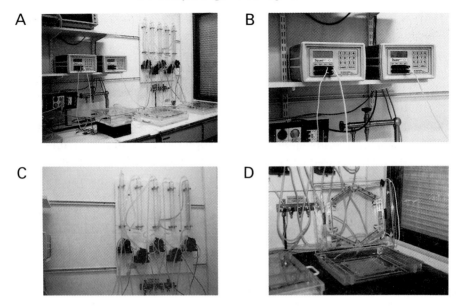

Figure 3 The PFGE system used in our laboratory. (A) Overview, two power supplies (*top left*) are used to drive several PFGE trays (*bottom*); two CHEF-boxes (*bench, middle* and *right*) and one 'Waltzer'-box (*left*) are visible. (B) Two four-channel Gene-tic power supplies. (C) Cooling system used. Five aquarium pumps circulate electrophoresis buffer through the refluxes (Section 4.1). (D) Close-up of a CHEF electrophoresis box. The lid has been opened to show the hexagonal electrode arrangement. Within the gel-box, the construction is visible that is used to fix the gel (four sets of pegs holding a removable table).

programmable output channels. Other commercially available power supplies and switch devices may lack some of the possibilities that will be described, but they are applicable with adaptations.

The Gene-tic allows each run to be subdivided into identical or non-identical cycles. Usually we run our gels in identical cycles of four to six hours. Each cycle consists of a linear or exponential time ramp of gradually increasing switch times. The shortest switch time, at the start of each cycle, defines the lower limit of separation; the longest switch time, at the end of each cycle, defines the upper limit of DNA molecules that are resolved. In the exponential time ramp the percentage of the total cycle duration at which 50% of the switch time increase is reached is set. A figure of 40% provides a time ramp which initially increases more rapidly. This setting improves the separation of larger DNA molecules. A short pause (a few percentage of the switch time) can be included each time when the electric field is reversed. This allows a relaxation of the DNA molecules and improves band sharpness above 400 kb. CHEF electrophoresis is frequently performed at a constant switch time, i.e. in one 'linear' cycle with no increase in switch time.

Temperature variations have a pronounced effect on the separation patterns obtained with PFGE. Since the high voltages used during electrophoresis generate a considerable amount of heat, a good cooling system is an essential component of a PFGE system. Electrophoresis buffer is circulated through the gel chamber by a simple aquarium pump, (e.g. type 1018 from EHEIM, Germany). Cooling is achieved by circulating the buffer through the inner coil of a glass reflux cooler using cooling water of 18 °C, controlled and circulated by a cryostat (*Figure 3C*). We find this set-up to be simple, cheap, and with much more heat exchange capacity than great lengths of rubber tubing immersed in cryostat chambers. Several glass coolers, each serving one electrophoresis tank, can be connected in series and cooled by one cryostat.

CHEF electrophoresis is performed in a rectangular gel-box with the electrodes fixed in a hexagonal configuration (*Figure 3D*, compare Chu *et al.* (24)) to the lid of the gel-box. The gel rests on a table and is kept in its place by two pegs at each corner. Our design allows the use of either 13 × 13 cm or 20 × 20 cm gels. Two removable bars separate the open buffer inlet and outlet spaces from the main gel chamber. A series of holes in each bar provides optimal cooling by guiding an even buffer stream just over the gel table. Samples can be loaded in one to three rows of lanes. When the gel is covered with a perspex plate, the gels can be stacked. The electric field is essentially homogeneous in the whole inner compartment, allowing the entire gel surface to be used. Thus, up to 60–100 samples can be run simultaneously in one gel, (e.g. for the analysis of YAC clones).

4.2 Gel preparation

Protocol 5. Gel preparation

1. Make up gels of 13.5 × 13.5 cm containing 120–150 ml 1% agarose (SeaKem FastLane, FMC) in 0.5 × TBE (45 mM Tris, 45 mM boric acid, 0.5 mM EDTA, pH 8.3).
2. Thoroughly remove the supernatant of the plugs containing digested DNA.
3. Transfer the plugs to slots pre-filled with buffer, using a Pasteur pipette bent into a hairpin and scalpel blade.
4. Similarly, transfer the plugs containing marker DNAs to the slots.
5. Equilibrate the gel with running buffer in the gel tray for 30–60 min. Electrophoresis is usually for 16–20 h, at 180 V, in 0.5 × TBE at 18 °C (see *Table 3*).

When necessary, the plugs can be fixed in the slots with agarose. Alternatively, the samples can be melted for 5 minutes at 65 °C and gently pipetted into the slots, using a cut-off yellow tip.

4.3 Planned DNA separations

The size range over which the DNA is separated is defined by the parameter settings of the program that drives the electrophoresis. In theory, specific combinations of these variables give a nearly infinite potential to separate DNA molecules of any size. After calibration of the system, specific DNA separations can be obtained by proper selection of these parameters (see *Figures 4* and *5* and *Table 3*). The preferred and best controllable parameters are switch time, voltage, agarose concentration, temperature, and reorientation angle.

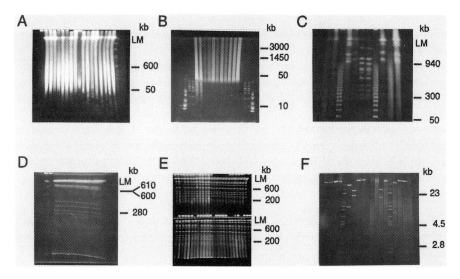

Figure 4 Examples of specific separations. Electrophoretic conditions and parameter setting used are given in *Table 3*. Fragment sizes are indicated in kb. (A) Standard 13 × 13 cm PFGE gel used for long-range physical mapping. Human DNA was digested with *Nae*I (lane 2), *Sac*II (lane 3, 18), *Nru*I (lane 4, 14), *Sal*I (lane 5, 16), *Sfi*I (lane 6, 15), *Mlu*I (lane 7, 12), *Bss*HII (lane 8, 17), *Spo*I (lane 9), *Not*I (lane 11), and *Spl*I (lane 13). (B) 11 different human genomic DNAs digested with *Bgl*II. Separation was done up to 5 Mb to allow hybridization analysis with centromeric repeat probes (51). (C) Ethidium bromide stained gel showing separation from 50–3000 kb. (D) Anomalous electrophoretic behaviour (see Section 4.3) specific for CHEF gels utilized for preparative separation of a 610 kb YAC from two 600 kb endogenous yeast chromosomes. (E) 20 × 20 cm gel used to analyse 50 individual recombinant YAC clones obtained from a homologous recombination experiment (39). (F) CHEF gel used for standard electrophoresis in a physical mapping experiment of cosmids cAL9 (52). Digestions with *Mlu*I (lane 1, 11), *Not*I (lane 2, 12), *Sal*I (lane 3, 13), *Xho*I (lane 4, 14), *Sma*I (lane 6, 16), *Kpn*I (lane 7, 17), *Nru*I (lane 8, 18), and *Sac*II (lane 9, 19). Markers used are a mixture of lambda digested with *Pst*I and *Hind*III (lanes F5, F15), lambda oligomers (lanes A1, C3, C8), *S. cerevisiae* (lanes A19, B2, B3, B16, B17, C6, C7, D1), *K. lactis* (lanes B1, B18, C5, C9), *C. albicans* (lanes C4, C11), and *S. pombe* (lanes B19, C1, C2). LM = limiting mobility zone.

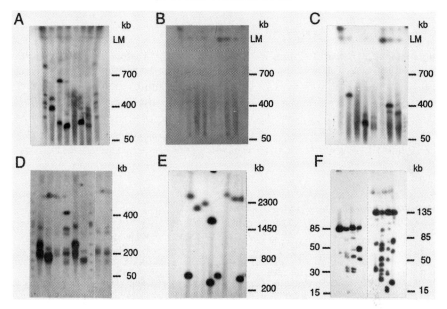

Figure 5 Hybridization analysis of PFGE gels. Electrophoretic conditions and parameter setting used are given in *Table 3*. Sizes are indicated in kb. Duo-blot of human genomic DNA hybridized with probes D4S139 (A) and Ho-1 (B). Digestions with *Nae*I (lane *A*7, *B*8), *Sac*II (lane *A*4, *B*7), *Nru*I (lane *A*1, *B*6), *Sal*I (lane *A*2, *B*5), *Sfi*I (lane *A*6, *B*4), *Mlu*I (lane *A*5, *B*3), *Bss*HII (lane *A*3, *B*2), *Spo*I (lane *A*8, *B*1), and *Not*I (lane 11). (C) Blot of *B* stripped and re-hybridized with cHD16-55. (D) Human genomic DNA isolated from blood leucocytes (lane 4, 9, 10) or EBV transformed cell lines (lane 2, 3, 5–7), digested with *Sfi*I and hybridized with pVK26.6 (membrane kindly provided by J. Saris). (E) Recombinant YAC clones separated overnight for size estimations in the range of 2300–3000 kb. (F) Partial digestion mapping of the YACs yDMD1 (*left*) and yDMD3 (*right*) (36). Restriction enzymes used were *Xho*I (lanes 1), *Xba*I (lanes 2), *Kpn*I (lanes 3), and *Sma*I (lanes 4). LM = limiting mobility zone.

i. Switch time

The switch time is defined as the interval between subsequent changes of the direction of the applied electric field. A longer switch time results in the resolution of larger DNA molecules (*Table 3*, compare also *Figure 4* panels F≫D≫A≫C≫B). Too long switch times combined with too high voltages will result in entrapment or breakage of the DNA molecules, a phenomenon visible by the apparent absence of a limiting mobility zone.

ii. Output voltage

The output voltage is given in V/cm over the gel. The larger the voltage applied, the faster the DNA migrates. Too large voltages result in the entrapment or breakage of the larger DNA molecules. Consequently, separations of increasingly larger molecules demand decreasing voltages and result in extended run times (compare *Figure 4*, panels *A*, *C*, and *B*). Furthermore, a

Table 3. Electrophoretic separation conditions

Separation range (kb)	V/cm	Run (h)	Switch time start	Switch time end	Agarose concentration	Figure
2.5–50	4.6	22	1	1	1%	4F
10–150	6.8	18	5	8	1%	5F
25–500[a]	5.4	16	1	30	1%	5D
25–630	6.8	18	33	33	1%	4D
50–950	5.7	20	40	80	1%	4E[b]
50–1100	6.8	20	40	80	1%	4A, 5A–C
50–1800	4.3	44	1	200	1%	4C
100–2000	2.7	68	100	650	1%	–
100–5000	2.3	96	300	1500	1%	4B
50–1000	4.6	20	1	90	0.7%[c]	–
100–3000	4.6	24	50	275	0.7%[c]	5E
100–5000	3.2	48	100	600	0.7%[c]	–

Conditions used to obtain PFGE separations within specific size ranges. All gels were run in the system described in 0.5 × TBE at 18°C and with a 2% pause for separations above 400 kb.
[a] FIGE gel run with four identical 4 hour cycles with 40% exponentially increasing switch times and 33% reversed electric field polarity.
[b] Gel measures 20 × 20 cm.
[c] SeaKem GOLD-agarose.
Switch times are given in seconds.

lower voltage requires longer switch times to obtain a similar separation (see *Table 3*).

iii. Agarose-gel
DNA mobility is influenced by the type of agarose used. Special types of agarose are available for PFGE applications. Decreasing agarose concentrations will increase DNA mobility but at the same time have the disadvantage of rendering gels difficult to handle. Recently developed agaroses, (e.g. Pronarose-D3, Hispanagar, Spain, and Gold-agarose, FMC, USA) have an extremely high gel strength and partly resolve this problem (see *Figure 5E*).

iv. Temperature
Higher temperature results in increased DNA mobility. Lower voltages and shorter switch times are possible when using higher temperatures. A disadvantage of temperatures above 20°C is that DNA degradation during electrophoresis also increases, especially for long run times, resulting in blurry bands.

v. Reorientation angle
In general, increasing angles between subsequent changes in the electric field direction result in a lower net DNA mobility. The influence of different angles on the separation patterns, although difficult to understand, is rather

profound. In most systems, the angle is fixed at about 120°C. In some systems it is varied to improve the separation of very large DNA molecules.

A very useful feature of PFGE systems is the near-linearity of the separation, (e.g. see the separation of lambda oligomers in *Figures 4C* and *5D*). This allows excellent resolution over a large size range, actually increasing near the top of the gel. Length differences of 20 kb are detected with equal ease between fragments ranging in size from 100 to 1000 kb. This makes the technique very powerful for the detection of minor deletions or rearrangements at great distances from the probes used (12, 33).

Near the top of the gel a zone of anomalous electrophoretic behaviour is observed in which chromosomes of different sizes are not resolved. This zone is called the 'limiting mobility' or 'compression' zone. In CHEF gels, the separation just below this zone is considerably extended. This feature can be used to separate DNA molecules that have a very small size difference; *Figure 4D* shows clear separation between the normally unresolved 600 and 610 kb chromosomes that have just left the limiting mobility zone.

4.4 Blotting of PFGE gels

Blotting of large DNA molecules separated on PFGE gels requires efficient DNA transfer. Acid depurination in our hands gives variable results and poor sensitivity. Very precise control of incubation times may reduce this effect in the hands of the individual experimenter. The most reliable and convenient method of reducing the DNA size and obtaining a complete transfer by conventional blotting, is by UV irradiation of the ethidium bromide stained gels.

Protocol 6. Blotting of PFGE gels

1. Stain the gel for 0.5–1 h in 2 μg/ml ethidium bromide.

2. Photograph immediately or destain by washing extensively in water to improve the contrast.

3. Reduce the DNA size by UV irradiation of the gel (upside down) at 180 000 μJ/cm^2 in the Stratalinker (Stratagene).

4. Wash the gel 2 × 15 min in 0.4 M NaOH. Rinse with water and blot the gel, upside down, overnight in 10 × SSC (150 mM NaCl, 15 mM Na citrate) to Hybond-N$^+$ or BioTrace^{RP} membrane.

5. Blot (at least) overnight. Change the paper towels in the morning, 1 h before dismounting the blotting-stack.

6. Soak the membrane (DNA side up) for 5 min on a puddle of 0.4 M NaOH. Wash 15 min in 2 × SSC, 0.2 M Tris, pH 7.5. Dry the blot for 1 h at 65°C.

4.5 Hybridization of PFGE gels

Hybridization of PFGE gels can be performed using standard hybridization protocols. The method that we prefer will be described below. In our hands, competitive DNA hybridization using whole cosmids reproducibly gives better results than the use of small genomic or cDNA probes.

Protocol 7. (Competitive) hybridization

1. Label 10 ng cosmid DNA. Purify over a Sephadex G-50 column in a Pasteur pipette.
2. Transfer half of the final eluate (about 200 μl) to an Eppendorf tube. Add 240 μl competitor DNA (500 ng/μl placenta DNA sonicated to 100–1000 bp).
3. Boil for 5 min and chill on ice. Add to a capped 10 ml plastic or polypropyllene tube with 1.5 ml hybridization buffer (0.125 M Na_2HPO_4 (pH 7.2 with H_3PO_4), 0.25 M NaCl, 1.0 mM EDTA, 7% SDS (Sigma, 44244), 10% PEG-6000 (BDH Chemicals Ltd., 44271), pre-heated to 65°C). Mix thoroughly, and incubate 90 min at 65°C in a water-bath (NB: time is critical!!).
4. Pre-hybridize the membranes in hybridization buffer for at least 10 min at 65°C in a water-bath.
5. Add the label to the pre-hybridized membranes, mix thoroughly, and hybridize overnight at 65°C.
6. Wash the membranes from 2 × SSC, 0.1% SDS (2 × 15 min), 1 × SSC, 0.1% SDS (2 × 15 min), down to 0.3 × SSC, 0.1% SDS (1 × 15 min) at 65°C.
7. Autoradiography takes four hours to three days at −70°C using an intensifying screen.

When single copy DNA probes are used, 10 ng labelled probe can be added directly to the pre-hybridized membranes (step **4**) after purification over a Sephadex G-50 column.

4.6 Re-hybridization of PFGE blots

Blots hybridized with a specific probe can be analysed with other probes after removal of the old hybridization signal ('stripping'). The method should be as gentle as possible for the DNA fixed to the membrane. We prefer to wash the filter for 30 minutes at 65°C in strip-mix (20 mM Na_2HPO_4 (pH 7.2 with H_3PO_4), 0.5 mM EDTA, 50% formamide, 0.5% SDS, 0.5 × SSC), followed by one wash for 15 minutes in 0.1 × SSC, 0.1% SDS, and air-drying of the filter. For security, stripped filters may be checked by overnight exposure.

5. PFGE applications

5.1 Analysis of chromosomal (human) DNA

For PFGE analysis of genomic eukaryotic DNA restriction enzymes have to be selected which cleave very infrequently. These enzymes recognize either octameric sequences, notably *Not*I and *Sfi*I (recognizing GCGGCCGC and GGCC(N)$_5$GGCC respectively), or hexanucleotides including one or more CpG dinucleotides since these sequences are underrepresented in the genome by an order of magnitude (34).

As expected, human genomic DNA cleaved with eight different rare-cutter enzymes (see *Table 2*) results in a digestion smear of fragments ranging from 50–10 000 kb (*Figure 4A*). Hybridization results depend primarily on the combination of enzyme and probe used. For new probes the results are, unfortunately, entirely unpredictable. *Figure 5* (panels *A*–*C*) shows the results when the blots of such gels are hybridized with different whole cosmid probes. Probe 1 gives a clear signal in every lane, probe 2 (from roughly the same genomic region) gives no discernible signal, and probe 3 gives a signal in only a few lanes. Within the resolution of the electrophoretic conditions chosen the results for probe 2, and for probe 3 with some enzymes, imply a total absence of sites in at least one direction. Thus, to exclude DNA degradation and lack of specific cleavage, new blots should be tested with a control hybridization, using a probe giving a known result with the specified enzymes before new probes are tested. A direct control for the digestion of rare-cutter sites can be obtained by a double digest with a frequently cutting enzyme, followed by hybridization of a control probe from the immediate vicinity of the site(s) for the rare-cutter(s) used, thus verifying double digestion of the fragment produced with the second enzyme alone (5).

Rare-cutter digestions often give multiple hybridizing DNA fragments (*Figure 5A*–*C*). Since over-digestion has no effect, this is not due to incomplete digestion but due to partial resistance of the sites to cleavage. The most likely cause of this resistance is partial methylation of the CpG dinucleotide(s) present in the recognition sequence of most rare-cutters. Consistent with its site lacking CpG, *Sfi*I mostly gives complete digestion patterns. Provided that the pattern becomes not too complex, partial digestion does extend the detection range of restriction sites since it permits the mapping of several adjacent sites (5, 11).

Since methylation is usually tissue-specific, it allows further physical mapping by comparison of the hybridization patterns obtained with DNA isolated from different sources (see *Figure 5D*). Moreover, methylation and underrepresentation of CpG in the genome are causally related (35), as are methylation and gene inactivation (34). Consequently, clustering of many rarecutter sites to a specific position on the physical map (a so called HTF island) is often indicative of the nearby location of (the 5′ end of) a gene. This at times reduces the resolution of multi-enzyme maps, i.e. when the sites of many of them are clustered.

The most general application of PFGE is the construction of physical maps in specific genomic regions (3–5, 11). To obtain a detailed physical map single and double digestions are performed with a large set of enzymes. The most complexing factor in this type of analysis is the sensitivity of the system to DNA overloading. Overloading complicates lane to lane comparisons and results in over-estimations of deduced fragment sizes since it decreases DNA mobility, (e.g. compare *Figure 5D* lanes 2 and 4). Overloading can be recognized by broadening and O-shaped appearance of the bands and affects the entire lane. Specific enzymes are also susceptible to local overloading namely in that part of the lane where the bulk of the DNA migrates, (e.g. for *Sfi*I the 100–300 kb range).

Once a good long-range physical map has been constructed for a specific genomic region, it can be used in further studies, e.g. to establish physical linkage between neighbouring loci or to study the presence of chromosomal rearrangements suspected to occur within or near a specific locus. A good example is the analysis of the Duchenne muscular dystrophy gene. At a time when the cDNA product of the DMD gene was not yet completely cloned, PFGE did allow the detection of large genomic deletions and duplications in over 50% of the patients (11, 33). Presently, PFGE analysis is especially valuable for studying the carrier status in DMD deletion families (10, 11, and C. Van Broeckhoven, personal communication).

5.2 YAC technology

The introduction of yeast artificial chromosome (YAC) cloning (14) has increased the size of clonable inserts to over 1 Mb. This quantum leap in cloning capacity has had a great impact on large-scale physical mapping projects, especially for the large and complex genomes of eukaryotes. YAC cloning allows the isolation of large DNA segments containing megabase genes or other large functional units consisting of several genes or entire gene families. Another benefit of the yeast system is that it facilitates the easy engineering of the cloned DNA to construct any desired piece of DNA for further studies.

PFGE techniques are essential for many procedures in YAC technology, for example

- analysis of the partially digested DNA prepared for subsequent YAC cloning
- preparative DNA size-selection preceding vector ligation (7, 8)
- analysis of size and stability of the YACs constructed

Additional applications include physical mapping (14, 36), preparative isolation of YAC DNA (37, 38), determination of YAC polarity (39), and analysis of meiotic recombinations between overlapping YACs (39, 40). The use of PFGE techniques for YAC analysis requires only minor adjustments of the established protocols for genomic DNA.

Figure 5F shows the power of PFGE as a tool for physical mapping of YACs. Total YAC clone DNA was digested with a dilution series for each enzyme used, resulting in a series of digests from very under-digested to almost complete. Subsequently, for each enzyme, a mixture is made of all degrees of digestion which is loaded into one lane of the gel. The gel was electrophoresed in such a way that an optimal resolution was obtained between 10 and 150 kb (36). After blotting and hybridization with a vector probe specific for one end of the YAC, a ladder-like restriction enzyme fingerprint is obtained displaying a detailed physical map of the entire YAC from one end inwards (14, 36). The deduced physical map should be perfectly co-linear with the genomic map and differences may only be observed when methylation sensitive enzymes are used. The discrepancies originate from methylation differences between the methylated genomic DNA and the unmethylated cloned YAC DNA, and in these cases the genomic fragments should be larger than cloned fragments.

Homologous recombination between partly overlapping YACs during yeast meiosis allows the generation of YACs spanning the entire genomic region and YACs containing only the overlapping region (39, 40). Initial selection of candidate recombinant clones can be easily achieved by PFGE sizing of the YACs (*Figure 5E*). In the experiment shown, the gel was loaded with several YACs obtained from a recombination between two YACs containing a segment of the Duchenne muscular dystrophy gene. Based on its increased size (compare with the parental YACs in lane 8), the only candidate large recombinant YAC is in lane 6. All other clones contain either parental YACs, the overlap recombinant, or YACs with internal rearrangements.

5.3 Further applications

The use of pulsed-field gel electrophoresis has undergone an amazing expansion, both in volume and in applications, since the first edition of this book was published in 1986. The analysis of eukaryotic genomic DNA and YACs described above represent only two major applications of a technique whose vast experimental potential is still limited by the creativity of its users. It is beyond the scope of this chapter to deal with all potential applications but below we will briefly describe some to illustrate the existing possibilities.

5.3.1 Two-dimensional electrophoresis

Physical mapping of large genomic regions can be enhanced considerably by the use of two-dimensional PFGE. In this technique, the DNA of interest is first digested with a rare-cutter restriction enzyme and separated on a PFGE gel. Subsequently, the entire lane is digested with a second enzyme and run on a second gel. After blotting, hybridization, and autoradiography complex megabase sized physical maps can be constructed from one experiment. Applications include the physical mapping of prokaryotic genomes (6), simple eukaryotic genomes, YACs, and also complex mammalian loci, e.g. containing entire gene families (41).

5.3.2 Preparative DNA isolations

PFGE can be adapted to a preparative use. Applications include the purification of or enrichment for DNA above a specific size range prior to YAC cloning (7, 8), and the enrichment of specific sequences from genomic cell hybrid DNA digests, preceding their subcloning in other vectors (9). Alternatively, the isolated DNA can first be amplified with the polymerase chain reaction (PCR) to amplify regions flanked by repetitive DNA sequences (42). A related application is to use Alu-PCR on gel slices obtained from PFGE gels loaded with different digests of DNA from specific cell hybrids to aid the rapid construction of a physical map (43).

5.3.3 Electrophoretic karyotyping

One of the first applications of PFGE has been the electrophoretic karyotyping of different yeast strains. Yeasts show considerable interstrain variation when the electrophoretic chromosome banding patterns are compared. This variability can not only be used to discriminate between different strains but also to assess their relatedness. In bacterial genetics, information of the electrophoretic karyotype is enhanced by restriction enzyme digestion of the bacterial DNA prior to electrophoresis. This application has great diagnostic potential for the identification of pathogenic strains of bacteria and mammalian parasites. Examples are veterinary pathogens, plasmodium, and trypanosome strain differentiation (16, 44–46). Recently, restriction endonuclease fingerprinting by PFGE was successfully used as an effective epidemiological diagnostic tool for the identification of *Staphylococcus aureus* strains (47).

5.3.4 10–50 kb range

Although PFGE is usually regarded as a technique to separate only large DNA molecules, i.e. above 100 kb, it can do much more than that. The technique has wide application for the separation of DNA molecules in a much smaller size range, extending the limits of conventional electrophoresis, for improved resolution of restriction fragment length polymorphisms (RFLPs) with fragments above 10 kb, and for partial digestion mapping of cosmids.

5.3.5 Molecular dynamics of DNA electrophoresis

The development of PFGE has revived interest in the basic principles behind the electrophoretic separation of DNA molecules considerably and showed that the existing theories about the mechanisms involved had to be revised. Direct observation of the movement of individual DNA molecules by fluorescent microscopy resulted in spectacular pictures, displaying for the first time high activity and complex motions of DNA molecules (19, 22).

5.3.6 DNA topology

Structural differences have a profound effect on the mobility of DNA molecules during pulsed-field electrophoresis. Circular molecules migrate differently from linear molecules, and supercoiled DNA has another mobility as relaxed DNA. PFGE can thus be used to study DNA topology, although the basic principles behind the mobility differences are still barely understood (48).

6. Future developments

Future advances on the separation of DNA molecules will focus mainly on the outer limits of the separation range, especially the upper size limit. The recent development of a system called 'secondary pulsed-field gel electrophoresis' (49) illustrates that particularly in the separation of megabase sized molecules many possibilties to improve the current resolution range are still unrecognized, and that an upper size limit is probably not yet reached. Furthermore, PFGE systems require the setting of many interdependent variables that all have their own greater or lesser influence on the electrophoretic patterns obtained. Most currently available PFGE systems and techniques were developed by trial and error and they give considerable variations when used in different laboratories. These aspects make it difficult for new users to set up a reliably working PFGE system. In our many contacts with end-users, we have recognized a strong need for simplified systems with limited options covering a moderate set of frequently used resolutions, (i.e. between 50–2000 kb). This area includes many diagnostic applications in hereditary and somatically acquired disease (mostly rearrangement syndromes and cancer). It is striking that this segment of the market has been largely neglected by the equipment development industry. The new trend to computerize laboratory equipment will certainly involve pulsed-field electrophoresis, since its complex nature constitutes a perfect target area for the further development of expert systems such as the Bio-Rad CHEF-DR II™.

Acknowledgements

We gratefully acknowledge the skilful assistance of L. Gerrese and R. D. Runia (Sylvius Laboratory, Leiden University) in the construction and modifications of the FIGE and CHEF boxes and the electronic equipment, and J. M. H. Verkerk and M. Rijnkels for their technical assistance with the setting up of the technology. This work was supported in part by grants from the Muscular Dystrophy Group of Great Britain and Northern Ireland, the Muscular Dystrophy Association of America, the Dutch Prevention Fund, and the Dutch Organization of Scientific Research (NWO).

References

1. Schwartz, D. C. and Cantor, C. R. (1984). *Cell,* **37,** 67.
2. Birren, B. W., Simon, M. I., and Lai, E. (1990). *Nucleic Acids Res.,* **18,** 1481.
3. Bottaro, A., de Marchi, M., Migone, N., and Carbonara, A. O. (1989). *Genomics,* **4,** 505.
4. Bucan, M., Zimmer, M., Whaley, W. L., Poustka, A.-M., Youngman, S., Allitto, B. A., Ormondroyd, E., Smith, B., Pohl, T. M., MacDonald, M., Bates, G. P., Richards, J., Volinia, S., Gilliam, T. C., Sedlacek, Z., Collins, F. S., Wasmuth, J. J., Shaw, D. J., Gusella, J. F., Frischauf, A.-M., and Lehrach, H. (1990). *Genomics,* **6,** 1.
5. Burmeister, M., Monaco, A. P., Gillard, E. F., van Ommen, G. J. B., Affara, N. A., Ferguson-Smith, M. A., Kunkel, L. M., and Lehrach, H. (1988). *Genomics,* **2,** 189.
6. Bautsch, W. (1988). *Nucleic Acids Res.,* **16,** 11461.
7. Anand, R., Villasante, A., and Tyler-Smith, C. (1989). *Nucleic Acids Res.,* **17,** 3425.
8. Larin, Z., Monaco, A. P., and Lehrach, H. (1991). *Proc. Natl Acad. Sci. U.S.A.,* **88,** 4123.
9. Anand, R., Honeycombe, J., Whittaker, P. A., Elder, J. K., and Southern, E. M. (1988). *Genomics,* **3,** 177.
10. Chen, J., Denton, M. J., Morgan, G., Pearn, J. H., and MacKinlay, A. G. (1988). *Am. J. Hum. Genet.,* **42,** 777.
11. den Dunnen, J. T., Bakker, E., van Ommen, G. J. B., and Pearson, P. L. (1989). *Br. Med. Bull.,* **45,** 644.
12. Cremers, F. P. M., Sankila, E.-M., Brunsmann, F., Jay, M., Jay, B., Wright, A., Pinckers, A. J. L. G., Schwartz, M., Van de Pol, D. J. R., Wieringa, B., De la Chapelle, A., Pawlowitzki, I. H., and Ropers, H. H. (1990). *Am. J. Hum. Genet.,* **47,** 622.
13. O'Reilly, M. A. J. and Kinnon, C. (1990). *J. Immunol. Meth.,* **131,** 1.
14. Burke, D. T., Carle, G. R., and Olson, M. V. (1987). *Science,* **236,** 806.
15. Albertsen, H. M., Abderrahim, H., Cann, H. M., Dausset, J., Le Paslier, D., and Cohen, D. (1990). *Proc. Natl Acad. Sci. U.S.A.,* **87,** 4256.
16. Robertson, J. A., Pyle, E. P., Stemke, G. W., and Finch, L. R. (1990). *Nucleic Acids Res.,* **18,** 1451.
17. Bernards, A., Kooter, J. M., Michels, P. A. M., Moberts, R. M. P., and Borst, P. (1986). *Gene,* **42,** 313.
18. van Ommen, G. J. B. and Verkerk, J. M. H. (1986). In *Human Genetic Diseases: A Practical Approach* (ed. K. E. Davies), pp. 113–33. IRL Press Ltd., Oxford.
19. Schwartz, D. C. and Koval, M. (1989). *Nature,* **338,** 520.
20. Deutsch, J. M. (1988). *Science,* **240,** 922.
21. Lalande, M., Noolandi, J., Turmel, C., Rousseau, J., and Slater, G. W. (1987). *Proc. Natl Acad. Sci. U.S.A.,* **84,** 8011.
22. Smith, S. B., Gurrieri, S., and Bustamente, C. (1990). In *Electrophoresis of large DNA molecules. Theory and application. Current Communications in Cell and Molecular Biology,* Vol. 1 (ed. E. Lai and B. W. Birren), pp. 55–79. Cold Spring Harbor Laboratory Press, Cold Spring Harbor.
23. Carle, G. R., Frank, M., and Olson, M. V. (1986). *Science,* **232,** 65.

24. Chu, G., Vollrath, D., and Davis, R. W. (1986). *Science,* **234,** 1582.
25. Stewart, G., Furst, A., and Avdalovic, N. (1988). *BioTechniques,* **6,** 68.
26. Kolble, K. and Sim, R. B. (1991). *Anal. Biochem.,* **192,** 32.
27. Southern, E. M., Anand, R., Brown, W. R. A., and Fletcher, D. S. (1987). *Nucleic Acids Res.,* **15,** 5925.
28. Ziegler, A., Geiger, K.-H., Ragoussis, J., and Szalay, G. (1987). *J. Clin. Chem. Clin. Biochem.,* **25,** 578.
29. Smith, C. L. and Cantor, C. R. (1987). *Methods in enzymology, Vol. 155; Recombinant DNA* (ed. R. Wu), pp. 449–67. Academic Press, New York.
30. Cook, P. R. (1984). *EMBO J.,* **3,** 1837.
31. Collins, F. S. and Weissman, S. M. (1984). *Proc. Natl Acad. Sci. U.S.A.,* **81,** 6812.
32. Hofker, M. H., Wapenaar, M. C., Goor, N., Bakker, E., van Ommen, G. J. B., and Pearson, P. L. (1985). *Hum. Genet.,* **70,** 148.
33. den Dunnen, J. T., Bakker, E., Klein-Breteler, E. G., Pearson, P. L., and van Ommen, G. J. B. (1987). *Nature,* **329,** 640.
34. Bird, A. P. (1986). *Nature,* **321,** 209.
35. Coulondre, C. Miller, J. H., Farabaugh, P. J., and Gilbert, W. (1978). *Nature,* **274,** 775.
36. Grootscholten, P. M., den Dunnen, J. T., Monaco, A. P., Anand, R., and van Ommen, G. J. B. (1991). *Technique,* **3,** 41.
37. Baxendale, S., Bates, G. P., MacDonald, M. E., Gusella, J. F., and Lehrach, H. (1991). *Nucleic Acids Res.,* **19,** 6651.
38. Couto, L. B., Spangler, E. A., and Rubin, E. M. (1989). *Nucleic Acids Res.,* **17,** 8010.
39. den Dunnen, J. T., Grootscholten, P. M., Dauwerse, J. D., Monaco, A. P., Walker, A. P., Butler, R., Anand, R., Coffey, A. J., Bentley, D. R., Steensma, H. Y., and van Ommen, G. J. B. (1992). *Hum. Mol. Genet.,* **1,** 19.
40. Green, E. D. and Olson, M. V. (1990). *Science,* **250,** 94.
41. Woolf, T., Lai, E., Kronenberg, M., and Hood, L. (1988). *Nucleic Acids Res.,* **16,** 3863.
42. Warren, S. T., Knight, S. J. L., Peters, J. F., Stayton, C. L., Consalez, G. G., and Zhang, F. (1990). *Proc. Natl Acad. Sci. U.S.A.,* **87,** 3856.
43. van Ommen, G. J. B., den Dunnen, J. T., Lehrach, H., and Postka, A. (1990). In *Electrophoresis of large DNA molecules. Theory and application. Current Communications in Cell and Molecular Biology,* Vol. 1 (ed. E. Lai and B. W. Birren), pp. 133–48. Cold Spring Harbor Laboratory Press, Cold Spring Harbor, NY.
44. Scherer, S. and Magee, P. T. (1990). *Microbiol. Rev.,* **54,** 226.
45. Corcoran, L. M., Thompson, J. K., Walliker, D., and Kemp, D. J. (1988). *Cell,* **53,** 807.
46. van der Ploeg, L. H. T., Schwartz, D. C., Cantor, C. R., and Borst, P. (1984). *Cell,* **37,** 77.
47. Goering, R. V. and Duensing, T. D. (1990). *J. Clin. Microbiol.,* **28,** 426.
48. Chu, G. (1992). *Electrophoresis,* **10,** 290.
49. Zhang, T. Y., Smith, C. L., and Cantor, C. R. (1991). *Nucleic Acids Res.,* **19,** 1291.
50. de Jonge, P., de Jongh, F. C. M., Meijers, R., Steensma, H. Y., and Scheffers, W. A. (1986). *Yeast,* **2,** 193.

51. Mahtani, M. M. and Willard, H. F. (1990). *Genomics,* **7,** 607.
52. Blonden, L. A. J., den Dunnen, J. T., van Paassen, H. M. B., Wapenaar, M. C., Grootscholten, P. M., Ginjaar, H. B., Bakker, E., Pearson, P. L., and van Ommen, G. J. B. (1989). *Nucleic Acids Res.,* **17,** 5611.

4

Fluorescent *in situ* hybridization

V. J. BUCKLE and K. A. RACK

1. Introduction

The hybridization of RNA *in situ* to the DNA of a cytological preparation was first described by Gall and Pardue in 1969 (1) and the general principles of the technique they developed remain the basis of contemporary studies. During the 1970s this procedure was used to localize repeated sequences within the human genome by autoradiographic detection of radioactively labelled probes, and during the 1980s the sensitivity of this approach was refined sufficiently to permit detection of unique sequences of only 500 base pairs in length. At the same time non-isotopic detection procedures were being developed, which have several advantages over autoradiography including speed, safety, and resolution of signal. Advances in this area in recent years have revolutionized the applications for *in situ* hybridization as a tool in genome research. In particular the use of competitive hybridization means an enormous size range of cloned DNA sequences are now available for use as probes, and fluorescent detection systems now permit the concurrent visualization of multiple probes. These powerful new techniques provide a tool not only for examining the location and distribution of sequences on metaphase chromosomes but also for investigating organization within the interphase nucleus.

This chapter provides details of current techniques for *in situ* hybridization employed in our laboratory and tries to touch on the enormous range of novel applications which place this methodology within easy reach of non-cytogeneticists.

2. Practical considerations

In situ hybridization of DNA to chromosome or interphase preparations requires several sequential procedures. These stages will be considered individually and variable practical parameters will be discussed, whilst our experimental procedures are provided separately as numbered protocols.

2.1 Chromosome and nuclei preparation

Chromosomes are prepared by standard techniques, using cell synchronization in order to obtain long chromosomes and maximum mitotic index where necessary. This can be achieved in two ways, both of which are detailed in *Protocol 1*. Addition of a high concentration of thymidine to a dividing population of cells will induce a block in the cell cycle, which can then be released and the chromosomes harvested when the synchronized cells reach early metaphase. Addition of 5-bromodeoxyuridine (BUdR) in place of thymidine has a similar effect but the resultant incorporation of BUdR into the DNA during early S phase (2) can be utilized to produce a replication banding pattern on the chromosomes after hybridization of the probe (Section 2.7).

Protocol 1. Preparation of chromosomes

1. To prepare phytohaemagglutinin (PHA) stimulated lymphocytes add 0.4 ml of whole blood to each culture flask containing 10 ml of RPMI culture medium (see *Table 1*) with PHA, and incubate for 72 h at 37°C. Where lymphoblastoid cell lines are used, grow until an healthy dividing population is obtained, with around 5×10^6 cells per 10 ml of culture medium.

2. Add either thymidine to a final concentration of 300 μg/ml or BUdR to a final concentration of 200 μg/ml of medium and incubate the culture for 16–17 h. During this time the cell cycle becomes blocked in S phase and the cells are synchronized at that point.

3. Wash the cells twice by centrifuging at 1000 *g* for 5 min and resuspend in fresh medium, in order to release the block in the cell cycle.

4. For the thymidine-synchronized cultures continue to culture for a further 5 h prior to harvesting. The BUdR-synchronized cells require excess thymidine at 10^5 M to be added to the fresh medium followed by a further culture time of 6–7 h. Colcemid is unnecessary as the cell population is now synchronized.

5. Centrifuge the cells as before, discard the supernatant, and resuspend in a hypotonic solution of 0.56% KCl, previously warmed to 37°C, and leave at that temperature for 10 min. Handle lymphoblastoid cells very gently from this point, no vortexing, since the metaphases are easily disrupted and the chromosomes become overspread after slide-making. An alternative hypotonic solution of 0.419% KCl can be used when chromosomes spread poorly.

6. Centrifuge, discard the supernatant, and resuspend the cells in a freshly prepared fixative of three parts AnalaR methanol to one part glacial acetic acid at 4°C. Add the fixative a drop at a time up to 10 ml, mixing thoroughly. Leave for 20 min at 4°C, then wash a further twice with fresh fixative.

7. Slides should be cleaned before use. Soak in strong detergent, (e.g. Lipsol, Decon) overnight, then rinse in distilled water, and soak in a weak

HCl solution for 1–2 h. Rinse thoroughly in distilled water, soak in ethanol, and air-dry. We also clean in this way the coverslips which are used for hybridization.

8. Place a drop of cell suspension on each glass slide and allow to air-dry. Check the quality of the preparation under phase contrast. Slides can be used the day after they are made but for long-term storage the slides can be kept in a sealed container with desiccant at −20°C. We have successfully used slides which have been kept in this way for three years.

9. For BUdR-synchronized preparations stain a test slide to ensure that the BUdR has incorporated correctly. Immerse in a solution of Hoechst 33258 diluted to 2 μg/ml in 2 × SSC for 30 min. Rinse in 2 × SSC. Place the Petri dish 20 cm from a long wave ultraviolet light source (Sylvania; Blacklite-blue) for 1 h. Stain the slide for 3 min in 10% Gurr's Improved Giemsa (BDH) diluted in pH 6.8 phosphate buffer. Rinse in the same buffer and air-dry.

Table 1. Composition of media and buffers

RPMI/PHA medium
100 ml	RPMI
2 ml	phytohaemagglutinin M-form
2000 U	sodium heparin
20 ml	fetal calf serum
200 mM	L-glutamine
10 000 U	penicillin
10 mg	streptomycin

DNase I dilution buffer
20 mM	$NaOOCCH_3$ pH 5.0
0.15 M	NaCl
50%	glycerol

2 × PCR buffer
10 mM	$MgCl_2$
100 mM	KCl
20 mM	Tris–HCl pH 8.4
0.2 mg/ml	gelatin

Hybridization mix
500 μl	formamide AR (Fluka)
100 μl	50 × Denhardt's solution
400 μl	25% dextran sulphate in 12.5 × SSPE

12.5 × SSPE
2.25 M	NaCl
125 mM	NaH_2PO_4
12.5 mM	Na EDTA pH 7.2

Many applications now make use of interphase nuclei, which are best harvested as a synchronized population in the G1 phase of the cell cycle when a hybridized probe will appear as a single dot along the chromatin fibre. After replication, hybridization signals will appear as paired dots which may cause confusion during analysis. Interphase cells in G1 can be obtained from cell lines if harvested when the cells are at complete confluency, all nuclei appear to be of similar size and no mitotic cells are seen.

2.2 Pre-hybridization procedures

The standard aceto-methanol fixative will remove basic proteins which might otherwise block efficient hybridization, however further fixation steps do appear to control background levels of hybridization. A 'post-fixation' step using formaldehyde is standard in our protocol and other laboratories find a ten minute fixation in acetone beneficial. We do still use RNase to strip native RNA, which can interfere with hybridization, from the chromosome preparations, although other laboratories have found this step to be unnecessary.

Occasionally the retention of cytoplasm around nuclei or chromosomes in some cell lines may interfere with optimal hybridization and a step using proteinase K or detergent to permeabilize the cells may be required (3).

Throughout these procedures slides should be handled with care and moved gently from one solution to another, in a glass rack wherever possible, in order to prevent loss of material from the surface of the slides.

Protocol 2. Treatment of slides prior to hybridization

1. Prepare a 10 mg/ml stock solution of RNase A for use by boiling for 10 min to remove any contaminating DNase, cool slowly to room temperature, and store frozen until required. Use at a concentration of 100 μg/ml in 2 × SSC. Place 100 μl of this solution upon each slide under a coverslip, and incubate in a moist chamber at 37°C for 1 h. A plastic sandwich box with a sheet of filter paper, moistened in 2 × SSC can be used for this purpose. Dip the slides in 2 × SSC to allow the coverslips to float off and wash in two changes of 2 × SSC.

2. Post-fix the slides by standing for 5 min in PBS/50 mM $MgCl_2$ followed by 10 min in PBS/50 mM $MgCl_2$/1% formaldehyde. Wash with PBS, dehydrate through an alcohol series (30 sec in each of 10%, 50%, 70%, 90%, and 100% ethanol), and air-dry. The slides can then be stored desiccated at 4°C for up to four weeks before use.

3. Denature the chromosomal DNA by incubating the slides in 70% formamide/0.1 mM EDTA in 2 × SSC, for 5 min at 70°C. Wash the slides in cold (4°C) 2 × SSC, followed by a further two washes in 2 × SSC, then dehydrate through an alcohol series as before.

2.3 DNA resources for use as probes

At the crudest level total human DNA can be used to characterize the human content of somatic cell hybrids although this does not provide information as to which chromosomes are present. Locus specific repeated sequences such as the alphoid centromere repeats and the heterochromatic repeats are particularly useful for tagging a particular chromosome of interest, or for rapid sexing, or scoring for chromosome aneuploidy. Ready biotinylated probes of this sort can be purchased commercially (Oncor). Unique sequences can be localized against normal or abnormal chromosomes. The minimum size limit for reliable detection is around three to five kilobases although there are reports of detection of unique sequence probes under one kilobase in length (4).

The latter two types of sequence avoid the problem of background hybridization because of a lack of interspersed repeats which are likely to recognize sites for hybridization on many other chromosomes. However such hybridization can be suppressed in a pre-hybridization reassociation with either total human DNA or Cot1 DNA. Whilst Cot1 DNA is more convenient to use as a competitor (15 minute reassociation time compared to two hours for total human DNA), total human DNA can provide more efficient competition for certain types of probe, for example a whole chromosome paint for one of the acrocentric chromosomes which is likely to cross-hybridize to common sequences on the other acrocentric chromosomes. Such competitive *in situ* suppression (CISS) means that large segments of the genome can be used as specific probes, including DNA cloned into bacteriophage (9–23 kb) and cosmid (40 kb) vectors (*Figure 1a–c*), and yeast artificial chromosomes (YACs) (200 kb–1 Mb) (*Figure 1d*).

Chromosome painting—the delineation of whole or part of a chromosome with a fluorescent probe (5, 6)—again relies on suppression hybridization and can be achieved using a variety of alternative resources. Flow sorted chromosomes are now available as whole chromosome libraries cloned into phage, plasmid, and cosmid vectors from ATCC. Phage libraries will obviously be least efficient in terms of vector-to-insert ratio. The libraries can be grown up and labelled in-house or purchased ready labelled (BRL, Oncor), although not all chromosomes are available yet commercially. Whole DNA prepared from somatic cell hybrids containing single human chromosomes has also been used for chromosome painting (7) but this again suffers from a high host-to-human DNA ratio. The development of a degenerate oligonucleotide primed polymerase chain reaction (DOP-PCR) (8) as a method of random amplification and labelling of DNA means that flow sorted chromosomes can be used directly as probes in place of whole chromosome libraries. In our experience the paint provided by a DOP-PCR labelling gives a fairly smooth coverage of the chromosome (*Figure 1e*), certainly superior to the banded pattern produced by an Alu-PCR labelling (9), and DOP-PCR paints are now

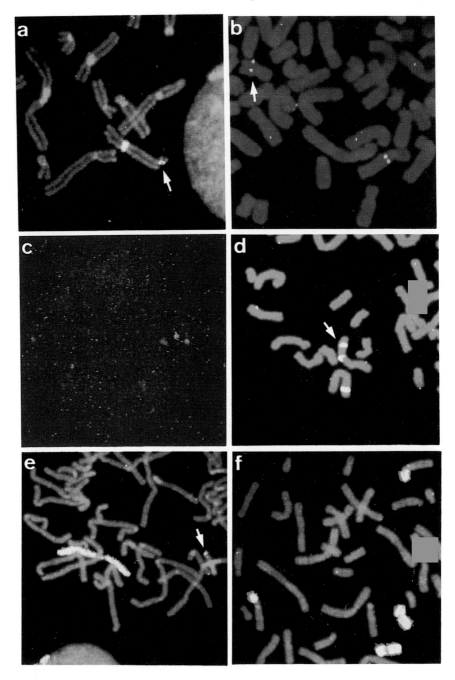

Figure 1 *In situ* hybridization and fluorescent detection. (a) Cosmid E2165 which maps distal to the FRAXA fragile site (*arrow*) on an X chromosome tagged with an alphoid repeat probe for the X centromere. (b) Cosmid 153 which maps at 5q13 and is retained on a deleted chromosome 5, del(5)(q13q33) (*arrow*), from the bone marrow cells of a patient with myelodysplasia. (c) Two biotinylated cosmids (M4 and 153) detected with Texas Red and one DIG labelled cosmid (JK53) detected with FITC, all derived from chromosome 5. JK53 maps in between the other two cosmids on normal interphase nuclei (30). (d) YAC 14E (insert size 430 kb) containing the EVI 1 gene which maps at 3q26 and crosses the breakpoint on an inverted chromosome 3, inv(3)(q21q26) (*arrow*), from a patient with acute myeloid leukaemia. (e) DOP-PCR paint generated from a flow sorted X chromosome. The pseudo-autosomal region on the Y chromosome short arm (2.5 Mb) is also labelled (*arrow*). (f) DOP-PCR paint generated from a flow sorted *de novo* 16p+ chromosome on normal chromosomes. The two normal chromosomes 16 are labelled and the extra material is derived from the short arm of chromosome 9. All images have been recorded on an MRC 600 confocal laser scanning microscope. (Acknowledgements with thanks to Georgina Flynn, Elaine Levy, and Angela MacCarthy for *Figures 1a, 1d,* and *1f* respectively.)

---

also commercially available (Cambio). The great advantage of using sorted chromosomes directly in this way is the facility for reverse chromosome painting where abnormal chromosomes can be sorted, labelled, and painted back on to normal chromosomes in order to determine their genomic derivation (10, 11) (*Figure 1f*). Microdissected chromosome regions have also been successfully labelled with the degenerate oligonucleotide primers and used as paints (12). With this approach selected regions of either normal or abnormal chromosomes can be dissected and labelled. Hence the method of choice for characterizing *de novo* chromosomes by reverse painting may to some extent depend on the availability of a FACSCAN or microdissection equipment. Somatic cell hybrid DNA labelled by Alu-PCR can also be used as a reverse paint in order to characterize the human content of the hybrid (13) (Section 2.4).

2.4 Probe labelling

An essential feature of this technique is a requirement for high quality DNA to ensure efficient labelling. Cloned DNA is best purified using a Qiagen column (Hybaid) or similar, or can be cleaned up using a CsCl ethidium bromide purification (14). For YACs, whole yeast DNA should be prepared using the guanidinium hydrochloride extraction procedure or a clean miniprep method (15), although an alternative protocol for labelling just the excised YAC band has also been reported (16).

One of the most critical factors for the success of *in situ* hybridization is the fragment size of labelled probe DNA in the hybridization mix. We aim for an average length of around 300 base pairs since larger fragments will interfere with proper penetration of the probe, and also produce high background signal on the slide. For this reason nick translation is a particularly suitable method of probe labelling since the end fragment size can be controlled by the amount of DNase I in the reaction mix. We make up our own labelling kit (*Protocol 3*) although commercial nick translation kits are available (Amersham, BRL). Fragment size can be checked by running out 100 ng of labelled probe on a 2% gel with suitable size markers (ΦX174 *Hae*III). The DNA should run as a smear stretching roughly between 100 bp and 500 bp. The precise amount of DNase 1 required to achieve this result will have to be titrated for each DNase 1 stock. Extra DNase 1 can always be added to the reaction mix if the average probe fragment size is too large, although in-efficient cutting generally implies that the probe DNA needs to be cleaned up.

Various methods can be used to label DNA non-isotopically. The most commonly used labels are biotin and digoxigenin which are purchased as dUTPs and incorporated by nick translation, as is dinitrophenol (17). DNA probes which have been chemically modified by acetylaminofluorine (18), sulphonation using a Chemiprobe kit (19), or mercuration (20) have also provided successful signals. More recently, dUTPs have become commercially available which have been directly coupled to fluorochromes (FITC, rhoda-mine, resorufin, coumarin) (Amersham, BRL) and the enormous advantage here is that with efficient labelling no detection steps are necessary in order to visualize the probe after hybridization, although of course the signal may well be weaker. Probes labelled with digoxigenin, biotin, or fluorochromes are stable at −20°C, therefore large amounts of DNA can be labelled in one reaction and stored frozen for future use.

Protocol 3. Nick translation

1. To 1 μg of probe DNA add:
- 1 μl 1 mM DIG-11-dUTP or biotin-16-dUTP (BRL)
- 5 μl 10 × nick buffer (0.5 M Tris–HCl pH 7.5, 50 mM MgCl$_2$, 0.5 mg/ml nuclease-free BSA)
- 5 μl 100 mM dithiothreitol
- 5 μl dNTP mix (0.5 mM each dATP, dCTP, dGTP, and 0.1 mM dTTP)
- 3 μl 3.5 U/μl DNA polymerase I
- 3 μl DNase I (1 ng/μl) (see *Table 1* for dilution buffer)
- sterile distilled water to a final volume of 50 μl

All the nick-kit stock solutions are stored at −20°C. Incubate the reaction mix for 2 h at 16°C, then stop the reaction with 10 μl 0.25 M EDTA, vortex, and place on ice.

2. Remove the unincorporated nucleotides by purifying the probe on a Select B column (CP Laboratories) according to the manufacturer's instructions.

3. Precipitate the probe in the presence of 50 μg *E. coli* tRNA and 50 μg salmon sperm DNA (200–500 bp) with 0.1 volume 3 M sodium acetate pH 5.6 and 2 volumes ethanol at −20°C for 2 h or overnight. Spin the DNA in a microcentrifuge at 4°C for 15 min. Tip off the supernatant and either air-dry the probe or dry in a vacuum pump. Resuspend the probe to an appropriate concentration in dH_2O and store at −20°C.

The polymerase chain reaction has proved a useful approach to general amplification and labelling of large segments of chromosomes, using primers for Alu or Kpn interspersed repetitive elements (IRS-PCR) or the degenerate oligonucleotide (DOP-PCR) mentioned above. Optimized PCR primers for consensus sequences within Alu repeats have been designed; these are species-specific and are therefore particularly suitable for defining the human content of somatic cell hybrids in a reverse paint and for labelling YACs (21–23). Otherwise the degenerate oligonucleotide generates a more even paint since it does not rely on the genomic distribution of a particular repeated sequence. DOP-PCR can be used to label flow-sorted chromosomes (10, 11) and microdissected regions (12).

Protocol 4. DOP-PCR amplification and labelling of flow sorted chromosomes

1. Combine in a sterile 1.5 ml microcentrifuge tube:

- 500 flow sorted chromosomes
- 50 μl 2 × PCR buffer (*Table 1*)
- 10 μl dNTP mix (2 mM each dATP, dCTP, dGTP, dTTP)
- 6.6 μl primer DOP 1 (5′ CCG ACT CGA GNN NNN NAT GTG G3′)
- 1.25 U *Taq* I polymerase
- sterile distilled water to a final volume of 100 μl

Mix, spin the reagents gently so as not to pellet the chromosomes, and overlay the reaction mix with 50 μl of mineral oil.

All reagents should be autoclaved and kept sterile. The use of sterile aerosol resistant pipette tips (Stratagene) will minimize contamination of the reaction mix. To achieve an optimum yield of PCR products the concentration of magnesium in the PCR buffer may need to be varied— we typically use a 5 mM final concentration.

2. Prepare a positive control (2.5 pg genomic DNA), and a negative control (no DNA) in the same manner.

3. Place in a PCR thermal cycler programmed with the following amplification routine:

Protocol 4. *Continued*

- 10 min at 93°C for denaturation
- 5 cycles of: 1 min at 94°C
  1.5 min at 30°C
  3 min at 30–72°C transition
  3 min at 72°C
- 35 cycles of: 1 min at 94°C
  1 min at 62°C
  3 min at 72°C

with an addition of 1 sec/cycle and the final extension step lengthened to 10 min.

Remove the oil and store the amplified reaction products at −20°C.

4. Take a 10 µl aliquot from these products, and from the positive and negative controls, and run on a 2% gel with the ΦX markers in order to check the success of the amplification and the size of the amplified products. They will normally run as a smear spanning the ΦX ladder.

5. Take a 5 µl aliquot from the reaction products into a new sterile tube, and make up a biotinylation reaction in the same manner as step **1** but with half quantities to a final volume of 50 µl, and with the addition of 5 µl 3 mM biotin-11-dUTP (Sigma).

6. Place in the PCR thermal cycler for 25 cycles at 62°C.

7. Remove the oil, purify the amplified DNA through a spin column (Select-B), and assess the DNA concentration (normally 20–50 ng/µl). Precipitate and redissolve in sterile water to a suitable concentration (see *Protocol 3*).

The potentials of oligonucleotide primed *in situ* DNA synthesis (PRINS) (24) do not yet appear to have been fully explored. This technique involves annealing a specific oligonucleotide on to chromosomes and then incorporating biotin-dUTP into an extension using DNA polymerase. This may prove a suitable approach to mapping short sequences.

2.5 Hybridization

The optimal DNA/DNA reassociation temperature (T_r) occurs some 25°C below the melting point (T_m) of the corresponding native duplex and stringent conditions will favour accurate base pairing. The presence of formamide will lower the T_r and so help maintain chromosome morphology. We continue to use a 5 × salt hybridization mix at 42°C since comparison of the signals with a seven kilobase unique sequence probe indicated no difference in efficiency between this and the more commonly used 2 × salt concentration at 37°C. Hybridization times depend on the vector to insert ratio of the probe; nick translated whole plasmids, phage, and cosmids, and chromosome libraries need only be left to hybridize overnight, as do PCR products, whilst whole yeast DNA requires up to four days (*Protocol 5, Table 2*). The area of slide to

Table 2. Hybridization parameters

Probe	Probe DNA concentration per slide (10 μl hyb mix)	Cot 1 competitor DNA concentration per slide	Pre-annealing time (minutes)	Hybridization time
Phage	200 ng	2.5 μg	15	O/N
Cosmid	80 ng	2.5 μg	15	O/N
YAC	300 ng	7.5 μg	15	~4 days
Centromeres	15 ng			O/N
Uniques	200 ng			O/N
Total human	50 ng			O/N
Chromosome library	100 ng	6.25 μg	15	O/N
FSC probes	100 ng	6.25 μg	15	O/N

be hybridized is quite small (18 × 18 mm) and it is possible to hybridize two probes side by side under sealed coverslips on one slide.

Protocol 5. Hybridization and subsequent washes

1. Dry down the appropriate concentration of probe and competitor (if required) in an Eppendorf using a vacuum pump and resuspend in 10 μl of hybridization mix per slide (see *Table 1*). The concentration of probe and competitor used in the hybridization will vary depending on the type of probe used (see *Table 2*). Probes in frequent use can be stored in hybridization mix at 4°C and can be repeatedly denatured and pre-annealed.

2. Denature the probe and competitor at 95°C for 10 min, chill on ice, and spin down briefly in a microcentrifuge. Pre-anneal at 37°C for 15 min if Cot1 DNA competition is used, or for 2 h if total genomic DNA is used.

3. Place 10 μl of probe hybridization mix on each slide and using forceps slowly lower an 18 × 18 mm coverslip from one side taking care to avoid leaving any bubbles. Seal with rubber solution to prevent evaporation of the hybridization fluid, and incubate for the appropriate length of time (see *Table 2*) at 42°C in the same sandwich boxes as described for the RNase step floating in a water-bath.

4. After hybridization, gently peel off the rubber solution without disturbing the coverslips. Remove these carefully by soaking the slides upright in 5 × SSC for some minutes until the coverslips float off easily. Transfer the slides to glass slide racks and wash in three changes of 2 × SSC at room temperature for 3 min. The stringency of subsequent washes to some extent depends on the nature of the probe and if problems of background signal are encountered then wash at a lower salt concentration. Wash the slides twice for 20 min each in 2 × SSC at 65°C, and follow with a 5 min wash in 2 × SSC at room temperature.

2.6 Signal detection

A variety of approaches to the visualization of non-isotopically labelled probes have been reported over the past few years. These include enzymatic detection with a coloured end precipitate (25), chemiluminescence with photobiotin, immunocytochemical detection with colloidal gold (26), and fluorochrome-conjugated antibodies with a sandwich amplification technique (5).

Fluorescent *in situ* hybridization is now the method of choice in most laboratories because of the speed, signal definition, and flexibility of this approach, and particularly in view of the developing possibilities for multi-colour labelling (27). Dual labelling of probes with biotin and digoxigenin has already proved useful for ordering sequences on chromosomes (28) and in interphase nuclei (29). A dual colour detection protocol can be used for two probes (one labelled with biotin and one with digoxigenin) but will also recognize a third probe labelled with both. Now with ratio labelling up to seven different sites can be differentiated with only the two labels although this does then require suitable software to interpret the images and assign pseudocolours (30). The single detection protocols given below (*Protocols 6 and 7*) will visualize the probe as a green signal against red chromosomes or nuclei. *Protocol 8* will produce a red signal for biotin labelled probes and a green signal for digoxigenin labelled probes, so the preparation must be counterstained with a dye which fluoresces at a different wavelength such as diamidino-phenylindole (DAPI).

All antibody incubation steps are carried out in a moist chamber. A slide rack placed on its side in a covered glass beaker containing moist tissue paper works well for this. Incubations for biotin detection are undertaken at room temperature whilst those for digoxigenin and dual detections are performed at 37°C and here the slides should be inverted in the slide rack to prevent drying out. All washing steps in these protocols are performed gently on a shaking table since this has a dramatic effect on levels of background signal.

Protocol 6. Biotin detection

1. Prepare a blocking solution of 5% non-fat dried milk (NFDM) in $4 \times$ SSC. Dissolve the powdered milk in the $4 \times$ SSC solution, then aliquot 1.5 ml into Eppendorfs and spin in a microcentrifuge for 10–15 min. Remove 1 ml of the clear supernatant from each aliquot into clean Eppendorfs, taking care not to disturb the pellet or remove any fat lying on the surface.

2. Make up the antibody dilutions in this blocking solution. Before use the antibody dilutions are spun in a microcentrifuge for 2 min. The antibody stocks in use are stored at 4°C for two weeks; for long-term storage aliquots are kept at −20°C.

3. Prepare a washing solution of $4 \times$ SSC/0.05% Tween20 and wash the slides in this for 3 min at room temperature.

4. Pre-block the slides prior to detection by placing 100 µl of the prepared blocking solution on to each slide, carefully lower a 24×50 mm coverslip on to each slide, and incubate in a moist chamber for 10 min. Wash the coverslips off in $4 \times$ SSC/0.05% Tween20, and drain the excess liquid from the slides but do not allow to dry.

5. Place 100 µl avidin-FITC antibody (5 µg/ml) (Vector DCS grade) on to each slide. Incubate under a coverslip for 20 min. Gently wash the slides in three changes of $4 \times$ SSC/0.05% Tween20 for 3 min each on a shaking table.

6. Place 100 µl biotinylated goat anti-avidin antibody (5 µg/ml) (Vector) on to each slide and incubate as before for 20 min. Gently wash the slides in three changes of $4 \times$ SSC/0.05% Tween20 for 3 min each on a shaking table.

Protocol 6. *Continued*

7. Place 100 µl avidin-FITC antibody (5 µg/ml) on to each slide and incubate for 20 min as before. Wash twice in 4 × SSC/0.05% Tween20 for 3 min and give a final wash in PBS for 3 min.

8. Drain the slides and mount in 30 µl antifade mounting medium (Vectashield H1000) containing 1 µg/ml propidium iodide (PI) counterstain under a 24 × 50 mm coverslip. We use PI since both counterstain and signal can be seen together under blue fluorescence without the use of additional filter sets. The concentration of PI in the antifade may need to be varied for optimum results.

9. Successive layers of biotinylated goat anti-avidin and avidin-FITC antibodies can be used to further amplify the signal. To remove the coverslips soak the slides in ethanol, rinse in fresh ethanol, and wash in 4 × SSC/ 0.05% Tween20 before applying the next antibody.

Protocol 7. Digoxigenin detection

1. Prepare a blocking solution of 0.5% non-fat dried milk (NFDM) in antibody buffer (Ab) (0.1 M Tris–HCl, 0.15 M NaCl) in the manner described in *Protocol 6*.

2. Prepare antibody dilutions in this blocking solution and microcentrifuge before use for 2 min (see *Protocol 6*).

3. Prepare a washing solution of Ab/0.05% Tween20 and wash the slides in this for 3 min at room temperature.

4. Pre-block for 20 min prior to detection by placing 100 µl of Ab/0.5% NFDM blocking solution under a 24 × 50 mm coverslip, invert the slides, and incubate at 37 °C. Wash the coverslips off in AB/0.05% Tween20, drain the excess liquid off the slide but do not allow to dry.

5. Place 100 µl of a 1/666 dilution of monoclonal mouse anti-digoxigenin antibody (Sigma) on to each slide under a coverslip and incubate as before for 30 min. Gently wash the slides in three changes of Ab/0.05% Tween20 for 3 min each on a shaking table.

6. Place 100 µl of a 1/1000 dilution of FITC-conjugated rabbit anti-mouse antibody (Sigma) on to each slide under a coverslip and incubate for 30 min. Wash the slides in three changes of Ab/0.05% Tween20 for 3 min each on a shaking table.

7. Place 100 µl of a 1/1000 dilution of FITC-conjugated mouse anti-rabbit antibody (Sigma) on to each slide and incubate under a coverslip for 30 min. Wash the slides twice as in the previous step and give a final wash in PBS for 3 min.

8. Drain the slides and mount as in *Protocol 6*.
9. Successive layers of rabbit anti-mouse FITC and mouse anti-rabbit FITC antibodies can be used to further amplify the signal. Remove the coverslips as described in the previous protocol.

Protocol 8. Dual or two colour detection

1. Prepare two blocking solutions of 5% NFDM in 4 × SSC (*Protocol 6*) and 0.5% NFDM in antibody buffer (Ab) (*Protocol 7*).
2. Prepare all the antibody dilutions in the Ab/0.05% NFDM blocking solution and microcentrifuge before use for 2 min.
3. Prepare a washing solution of Ab/0.05% Tween20 and wash the slides for 3 min at room temperature.
4. Pre-block the slides in 100 μl of the 4 × SSC/0.5% NFDM blocking solution under a 24 × 50 mm coverslip for 10 min at room temperature. Wash the coverslip off in Ab/0.05% Tween20, drain the excess liquid off the slide but do not allow to dry.
5. Place 100 μl of the first layer of antibodies containing both a 8/1000 dilution of Avidin Texas Red (Vector) and a 1/666 dilution of monoclonal mouse anti-digoxigenin on to each slide under a 24 × 50 mm coverslip. Incubate the inverted slides for 30 min. Gently wash the slides in three changes of Ab/0.05% Tween20 for 3 min each on a shaking table.
6. Place 100 μl of the second antibody layer containing both a 1/100 dilution of biotinylated goat anti-avidin and a 1/1000 dilution of rabbit anti-mouse FITC on to each slide. Incubate under a coverslip for 30 min. Wash the slides in three changes of Ab/0.05% Tween20 for 3 min each on a shaking table.
7. Place 100 μl of the third antibody layer containing both a 8/1000 dilution of Avidin Texas Red and a 1/100 dilution of mouse anti-rabbit FITC on to each slide. Incubate under a coverslip for 30 min. Wash the slides twice as described in the previous step and give a final rinse in PBS for 3 min.
8. Drain the slides and mount in 30 μl of Vectashield antifade mountant containing 1.5 μg/ml DAPI counterstain under a 24 × 50 mm coverslip.
9. Successive application of antibody layers two and three will further enhance the signal. Remove the coverslips as described in *Protocol 6*.

2.7 Chromosome banding

The location of sequences by *in situ* hybridization directly on to banded chromosomes is a rapid approach to generating high resolution physical maps. The assignment of a sequence to an individual band also provides information on the general functional significance of that region since light G-bands are early replicating, rich in CpG islands, contain many housekeeping genes, and large numbers of Alu repeats, whilst dark G-bands are late replicating with fewer but tissue-specific genes and more Kpn repeats. Fractional length measurements of a signal along a chromosome can not realistically be related to band position and so the novel assignment of a sequence by FISH should be undertaken against a banding pattern of relevance to classical G-band nomenclature. Several such techniques have been reported which involve the incorporation of BUdR into chromosomes prior to harvesting, followed by detection with an anti-BUdR antibody (31), by counterstaining with DAPI and PI (32), or with Hoechst 33258 and PI (33). An excellent R-band pattern can be produced by incorporating Alu-PCR products into the hybridization mix (34) and G-bands have also been reported by counterstaining with PI at pH11 (35).

We routinely use a mixture of DAPI (1.5 µg/ml) and PI (0.75 µg/ml) counterstain in the antifade mountant. This produces a G-band pattern when the chromosomes are viewed under ultraviolet filters and an R-band pattern under the green filter set. The DAPI counterstain on its own (*Protocol 8*) will still produce a G-band pattern which can be useful after a dual colour detection.

2.8 Visualization of signal

Many of the basic applications for *in situ* hybridization can be undertaken with a standard fluorescence microscope provided that it is fitted with suitable filter sets. Photographs can be taken on Fujiichrome 400 set at 800 ASA (prints) or Scotch 3M 640T (slides) with an exposure time of the order of one minute. Double or triple exposures using different filter sets will be necessary for viewing the results of any dual colour detections and this process is not only time consuming but also, particularly for analysis involving fine measurements, may introduce errors through displaced registration of images. There are two current solutions to these problems:

- confocal microscopy
- multiple band pass filter sets

A confocal laser microscope operates by scanning the specimen at a given focal depth and converting the image into digital form through photomultiplier tubes. There are two laser lines which operate at wavelengths suitable for FITC and for PI/Texas Red/TRITC. The signals excited at the two different wavelengths are captured separately, then the two digital images can be merged with perfect registration. Accompanying software enables the images to be manipulated and facilitates rapid archiving by optical disc. Most importantly the sectioning capacity of the laser means that three-dimensional

FISH: fluoresant in situ hybridization

structure can be investigated which makes this a particularly suitable tool for the study of nuclear organization. The disadvantages of this equipment for molecular cytogenetic work is the restriction to only two wavelengths when so much more information could be derived from the concurrent availability of an ultraviolet wavelength. This disadvantage can be overcome and some of the advantages of confocal microscopy extended by the use of triple band pass filters (Omega) in conjunction with a cooled CCD camera (Photometrics). This type of camera operates with long integration times at low light intensity to produce a digital image with similar potential for image manipulation and will collect images at whatever wavelength it is exposed to.

2.9 Analysis of signal

Under optimum conditions the majority of metaphases will be informative after hybridization with YAC or cosmid probes, certainly over 80% of metaphases will exhibit signal on each chromatid of the target chromosome. For smaller probes this figure become reduced, for example to about 30% for a three kilobase unique sequence. Provided that background levels of hybridization are kept low and a signal is only scored when both chromatids are labelled, an unequivocal answer is easily obtained.

YAC probes can give quite a diffuse signal in interphase nuclei and for this reason are not suitable for experiments involving measurement of distance between signals. Relative measurements can be scored rapidly on archived digital images however it is perfectly possible to project colour slides taken down an ordinary microscope on to a wall and take measurements with pen and ruler. Statistical analysis will normally be necessary to look at the significance of this sort of data and also to calculate how many cells need be scored for a reliable interpretation.

2.10 Troubleshooting

Use a commercially available ready labelled repetitive probe such as a centromeric alphoid repeat in order to check the success of basic hybridization and detection procedures. If problems are encountered here then look at slide fixation, denaturation, and washing stringency, and check viability of the antibodies.

(a) Cells lost from slide. Insufficient fixation can lead to loss of material from the surface of the slide. Throughout the procedure slides should be handled with care especially when coverslips are being removed.

(b) Puffy chromosomes. The slides may have been denatured at too high a temperature.

(c) No signal. This may be due to insufficient probe DNA in the hybridization mix, inadequate denaturing of the probe or chromosomes, or probe fragment size being too small. The concentration of any new DNA to be used should be carefully checked by fluorimeter or on a gel against a range of concentrations of uncut lambda phage DNA. Fragment size of a labelled DNA stock can be checked by running against ΦX markers on a 2% gel.

(d) Background signal on chromosomes and nuclei. Insufficient competitor DNA will fail to block cross-hybridization to interspersed repetitive sequences. Some DNA probes may require additional competition to that suggested in *Table 2* because of the sequences they contain.

(e) Background signal all over the slide. When the labelled probe DNA fragment size is too large we see non-specific adherence to cytological material and also large sparkly accumulations of signal all over the slide. Re-check the quality of the DNA if it repeatedly cuts poorly. It is important that the number of antibody layers be kept to a minimum since amplification of the signal is generally accompanied by increased levels of background. Inadequate blocking of the slide with non-fat dried milk will also lead to high background.

3. Applications

The speed and flexibility of fluorescent *in situ* hybridization procedures now mean that this technique can be used not only in research but also as an aid to clinical diagnosis. In addition the ability to screen interphase nuclei does promise to overcome limitations imposed by the requirement of cytogenetic analysis for a dividing cell population which may be non-representative and difficult to obtain.

3.1 Direct localization within the genome

Any DNA sequence over approximately three kilobases (in plasmid, phage, cosmid, or YAC) can rapidly be given a *de novo* regional assignment. There are reports of successful hybridizations with probes of less than one kilobase (4) but our approach for such small probes would be to pull out a corresponding phage or cosmid. The best resolution for an assignment will be obtained where the hybridization signal can be seen directly against banded chromosomes, although large numbers of cosmids can be rapidly grouped along a chromosome using translocation breakpoints (36).

All YAC libraries suffer to a greater or lesser extent from the phenomenum of co-ligation, where a YAC is comprised of two or more non-contiguous fragments derived from different regions of the genome. This will become apparent when the YAC is mapped on normal chromosomes by FISH and can be important information, particularly in the construction of YAC contigs. It appears that YAC libraries with a larger average insert size have a higher frequency of co-ligation events.

3.2 Sequence order on the chromosome

There are a variety of approaches available for determining the linear order of DNA sequences along a chromosome, including genetic recombination, radiation or deletion hybrids, and pulsed-field gel electrophoresis. *In situ* hybridization combined with fluorescent multi-colour detection is a rapid method for ordering sequences over a wide range of distances. Sequences separated by over two to four million base pairs can be ordered on long chromosomes (28) but below that distance the signals either coalesce or sit side by side across the width of the chromatids. However as sequences become too close to be resolved at metaphase, information can still be obtained by analysis of interphase nuclei where the chromatin fibres are approximately 100-fold less condensed, which permits the resolution and ordering of sequences only 100 kb apart (37). Pronuclear interphase chromatin from human sperm can be induced to decondense to a further degree after fusion with hamster ova (38), and probes separated by only 40 kb have been consistently resolved on this material. The linear order appears to break down above one million base pairs because of folding of the chromatin. Techniques for preparing extended DNA fibres from interphase chromatin are being developed in order to improve further the limits of resolution at interphase (39, 40, 41).

To obtain information on linear order, whether at metaphase or inter-phase, we would use either two colours for three probes (red/biotin and green/DIG) (*Figure 1c*), or three colours where the third probe appears yellow because it is comprised of both biotin and DIG labelled DNA. By varying the ratio of biotin labelled to DIG labelled DNA for additional probes it is possible to increase the number of probes that can be differentiated in a single hybridization (42). For example, a hybridization scheme comprising:

- probe A = 100% biotin labelled DNA
- probe B = 80% biotin labelled +20% DIG labelled DNA
- probe C = 50% biotin labelled +50% DIG labelled DNA
- probe D = 20% biotin labelled +80% DIG labelled DNA
- probe E = 100% DIG labelled DNA

will distinguish five sites using only two labels. The intensity of signal from different labellings can vary quite considerably and this should be assessed for each probe so that the ratio of combined labelled DNA used may actually differ from the required signal intensity ratio. Analysis of this sort of data is best performed on a digital image where the relative intensities of signal at different wavelengths can be recorded and an appropriate pseudocolour assigned to that site.

A linear relationship has been established between genomic distance and the distance between FISH signals in interphase nuclei over the range 100 kb–1 Mb

(37), which means that some estimation of intergenic distances between sequences in this range can be attempted. Above that range interphase distances do remain consistent but increase less steeply than their genomic counterparts.

3.3 Structural chromosome abnormalities

Chromosome breakpoints associated with any cytogenetic abnormality (translocations, deletions, inversions, insertions, ring chromosomes, fragile sites) can be characterized by hybridization with probes mapping within the vicinity of the abnormality (43, 44) (*Figure 1a* and *b*). The use of FISH with YAC clones in particular is a valuable approach to positional cloning where chromosome breakpoints are available since strong signals can be observed on either side (*Figure 1d*). Once a YAC is identified as crossing the break-point, it can be subcloned as a first step to cloning the breakpoint region.

The high hybridization efficiency of large cloned fragments means that well-characterized chromosome translocations can be screened for in inter-phase nuclei (and also poor quality chromosome preparations which are unsuitable for standard banded analysis) in a diagnostic context. The Phila-delphia translocation can be studied in this way by hybridizing cosmid probes for 5'BCR and 3'ABL sequences using a two colour detection—the signals are juxtaposed where the translocation is present (45). This approach is particularly suited to work on leukaemias and solid tumours (46) where mitotic preparations may be difficult to obtain, may not be representative of the cell population as a whole, and can not provide information on cell lineage, and may prove an alternative approach to screening for minimal residual disease. A signal which becomes split rather than two signals brought together is an another option for interphase screening, by using a YAC which is known to cross a breakpoint.

3.4 Numerical chromosome abnormalities

Locus specific repeated sequences can be used to screen rapidly for chromo-some aneuploidy in either metaphase or interphase cells which as outlined above can be particularly informative in studies of malignant disorders. For interphase analysis care should be taken to compare results with a control cell population or another internal control since a percentage of aneuploid nuclei will always be scored if only for technical reasons. Interphase studies on monosomy 7 in acute myeloid leukaemia enabled karyotype to be correlated to differentiation status (47), and trisomy 12 has been shown to be a more com-mon finding in CLL than was judged from standard cytogenetic analysis (48).

3.5 Chromosome painting

Whole chromosomes or regions of chromosomes can be defined by chromo-some paints derived from a variety of sources (Section 2.3). Metaphase preparations can be screened for aneuploidy (6, 49) and for structural abnormalities, including induced damage (50, 51). Cancer cytogenetic studies are likely to make increasing use of this technique since information can be derived even from poor quality preparations and paints can be used in combination with standard probes to recognize particular chromosome segments after complex rearrangements. Routine cytogenetic laboratories are finding that chromosome painting is a useful adjunct to classical analysis of cryptic rearrangements (52), and in prenatal diagnosis (53), although some limitations have been encountered in terms of consistent coverage and sensitivity. Using DOP-PCR we find that an X chromosome paint is sufficiently sensitive to detect the pseudo-autosomal region of two and a half million base pairs on the Y chromosome short arm (*Figure 1e*). Without further contraction of the chromatin we have found the signal from whole chromosome paints in interphase cells to be too diffuse for reliable analysis.

Reverse chromosome painting which combines either flow sorting or microdissection with FISH is a rapid method for characterizing chromosome abnormalities (*Figure 1f*), and will permit correlation of clinical phenotype with aneuploidy for particular chromosome regions. Micro-FISH allows the rapid generation of region-specific fluorescent probes and again will prove very useful for cancer cytogenetics.

Somatic cell hybrids can be characterized by forward painting with a paint for the selectable chromosome, or by a reverse painting approach where the hybrid DNA is labelled by Alu-PCR and painted back on to normal chromosomes (13, 54) (see Section 2.4). This reverse technique will provide information on all the chromosomes and chromosome regions carried within that particular hybrid.

3.6 Nuclear organization

Investigations of three-dimensional organization within the interphase nucleus suggest that various nuclear functions such as transcription, replication, and repair may be compartmentalized just as cytoplasmic functions appear to be (55), and hence that gene activity may be controlled in part by nuclear position. FISH analysis in conjunction with confocal microscopy will provide a powerful approach to investigating the location of genes and larger chromosome regions in different functional states.

3.7 Other applications

Dual colour hybridization with X and Y chromosome-specific probes is a rapid approach to sexing metaphase or interphase cells. This promises to be a credible technique for the diagnosis of sex in human pre-implantation embryos (56), and has been used to determine the frequency of meiotic nondisjunction by sexing human spermatozoa (57).

The timing of replication during S phase for any given sequence has been shown to be reflected in an unsynchronized cell population by the type of FISH signal observed in interphase nuclei (58). Two singlet hybridization dots will predominantly be observed for early replicating sequences whereas late replicating sequences are represented predominantly by doublets. This observation has led to the characterization of replication time zones and means that correlations can be made between timing of replication and the functional state of any given gene.

Chromosome painting of meiotically dividing cells from testicular biopsies now permits a quantitative analysis of segregation in male carriers of structural rearrangements (59).

References

1. Gall, J. G. and Pardue, M. L. (1969). *Proc. Natl Acad. Sci. U.S.A.*, **63**, 378.
2. Lemieux, N., Drouin, R., and Richer, C.-L. (1990). *Hum. Genet.*, **85**, 261.
3. Manuelidis, L. (1985). *Focus*, **7**, 4.
4. Viegas-Pequignot, E., Berrard, S., Brice, A., Apiou, F., and Mallet, J. (1991). *Genomics*, **9**, 210.
5. Pinkel, D., Straume, T., and Gray, J. W. (1986). *Proc. Natl Acad. Sci. U.S.A.*, **83**, 2934.
6. Lichter, P., Cremer, T., Tang, C.-j. C., Watkins, P. C., and Manuelidis, L. (1988). *Proc. Natl Acad. Sci. U.S.A.*, **85**, 9664.
7. Landegent, J. E., Jansen in de Wal, N., Dirks, R. W., Baas, F., and van der Ploeg, M. (1987). *Hum. Genet.*, **77**, 366.
8. Telenius, H., Carter, N. P., Bebb, C. E., Nordenskjold, M., Ponder, B. A. J., and Tunnacliffe, A. (1992). *Genomics*, **13**, 718.
9. Suijkerbuijk, R. F., Matthopoulos, D., Kearney, L., Monard, S., Dhut, S., Cotter, F. E., Herbergs, J., van Kessel, A. G., and Young, B. D. (1992). *Genomics*, **13**, 355.
10. Carter, N. P., Ferguson-Smith, M. A., Perryman, M. T., Telenius, H., Pelmear, A. H., Leversha, M. A., Glancy, M. T., Wood, S. L., Cook, K., Dyson, H. M., Ferguson-Smith, M. E., and Willatt, L. R. (1992). *J. Med. Genet.*, **29**, 299.
11. Telenius, H., Pelmear, A. H., Tunnacliffe, A., Carter, N. P., Behmel, A., Ferguson-Smith, M. A., Nordenskjold, M., Pfragner, R., and Ponder, B. A. J. (1992). *Genes, Chromosomes, Cancer*, **4**, 257.
12. Meltzer, P. S., Guan, X.-Y., Burgess, A., and Trent, J. M. (1992). *Nature Genet.*, **1**, 24.
13. Lichter, P., Ledbetter, S. A., Ledbetter, D. H., and Ward, D. C. (1990). *Proc. Natl Acad. Sci. U.S.A.*, **87**, 6634.
14. Saunders, S. E. and Burke, J. F. (1990). *Nucleic Acids Res.*, **18**, 4948.
15. Borts, R. H., Lichten, M., and Haber, J. E. (1986). *Genetics*, **113**, 551.
16. Selleri, L., Hermanson, G. G., Eubanks, J. H., and Evans, G. A. (1991). *GATA*, **8**, 59.
17. Lichter, P., Tang, C. C., and Call, K. (1990). *Science*, **247**, 64.
18. Landegent, J. E., Jansen in de Wal, N., van Ommen, G.-J. B., Baas, F., de Vijlder, J. J. M., van Duijn, P., and van der Ploeg, M. (1985). *Nature*, **317**, 175.

19. Renz, M. and Kurz, C. (1984). *Nucleic Acids Res.,* **12,** 3435.
20. Hopman, A. H. N., Wiegant, J., Tesser, G. I., and van Duijn, P. (1986). *Nucleic Acids Res.,* **14,** 6471.
21. Tagle, D. A. and Collins, F. S. (1992). *Hum. Mol. Genet.,* **1,** 121.
22. Lengauer, C., Green, E. D., and Cremer, T. (1992). *Genomics,* **13,** 826.
23. Breen, M., Arveiler, B., Murray, I., Gosden, J. R., and Porteous, D. J. (1992). *Genomics,* **13,** 726.
24. Gosden, J., Hanratty, D., Starling, J., Fantes, J., Mitchell, A., and Porteous, D. (1991). *Cytogenet. Cell Genet.,* **57,** 100.
25. Garson, J. A., van den Berghe, J. A., and Kemshead, J. T. (1987). *Nucleic Acids Res.,* **15,** 4761.
26. Hutchison, N, J., Langer-Safer, P. R., Ward, D. C., and Hamkalo, B. A. (1982). *J. Cell. Biol.,* **95,** 609.
27. Nederlof, P. M., Robinson, D., Abuknesha, R., Wiegant, J., Hopman, A. H. N., Tanke, H. J., and Raap, A. K. (1989). *Cytometry,* **10,** 20.
28. Lichter, P., Tang, C.-j. C., Call, K., Hermanson, G., Evans, G. A., Housman, D., and Ward, D. C. (1990). *Science,* **247,** 64.
29. Morrison, K. E., Daniels, R. J., Suthers, G. K., Flynn, G. A., Francis, M. J., Buckle, V. J., and Davies, K. E. (1992). *Am. J. Hum. Genet.,* **50,** 520.
30. Ried, T., Baldini, A., Rand, T. C., and Ward, D. C. (1992). *Proc. Natl Acad. Sci. U.S.A.,* **89,** 1388.
31. Vogel, W., Autenrieth, M., and Mehnert, K. (1989). *Chromosoma (Berl.),* **98,** 335.
32. Fan, Y.-S., Davies, L. M., and Shows, T. B. (1990). *Proc. Natl Acad. Sci. U.S.A.,* **87,** 6223.
33. Cherif, D., Julier, C., Delattre, O., Derre, J., Lathrop, G. M., and Berger, R. (1990). *Proc. Natl Acad. Sci. U.S.A.,* **87,** 6639.
34. Baldini, A. and Ward, D. C. (1991). *Genomics,* **9,** 770.
35. Lemieux, N., Dutrillaux, B., and Vicgas-Pequignot, E. (1992). *Cytogenet. Cell Genet.,* **59,** 311.
36. Kievits, T., Dauwerse, J. G., Wiegant, J., Devilee, P., Breuning, P. H., Cornelisse, C. J., van Ommen, G.-J. B., and Pearson, P. L. (1990). *Cytogenet. Cell Genet.,* **53,** 134.
37. Lawrence, J. B., Singer, R. H., and McNeil, J. A. (1990). *Science,* **249,** 928.
38. Brandriff, B. F., Gordon, L. A., Tynan, K. T., Olsen, A. S., Mohrenweiser, H. W., Fertitta, A., Carrano, A. V., and Trask, B. J. (1992). *Genomics,* **12,** 773.
39. Houseal, T. W., Dackowski, W. R., Landes, G. M., and Klinger, K. W. (1992). *Am. J. Hum. Genet.,* **51,** A9.
40. Wiegant, J., Kalle, W., Mullenders, L., Brookes, S., Hoovers, J. M. N., Dauwerse, J. G., van Ommen, G. J. B., and Raap, A. K. (1992). *Hum. Mol. Genet.,* **1,** 587.
41. Heng, H. H. Q., Squire, J., and Tsui, L.-C. (1992). *Proc. Natl. Acad. Sci. U.S.A.,* **89,** 9509.
42. Dauwerse, J. G., Wiegant, J., Raap, A. K., Breuning, M. H., and van Ommen, G. J. B. (1992). *Hum. Mol. Genet.,* **1,** 593.
43. Dauwerse, J. G., Kievits, T., Beverstock, G. C., van der Keur, D., Smit, E., Wessels, H. W., Hagemeijer, A., Pearson, P. L., van Ommen, G.-J. B., and Breuning, M. H. (1990). *Cytogenet. Cell Genet.,* **53,** 126.

44. Alitalo, T., Willard, H. F., and de la Chapelle, A. (1989). *Cytogenet. Cell Genet.*, **50,** 49.
45. Arnoldus, E. P. J., Wiegant, J., Noordermeer, I. A., Wessels, J. W., Beverstock, G. C., Grosveld, G. C., van der Ploeg, M., and Raap, A. K. (1990). *Cytogenet. Cell Genet.*, **54,** 108.
46. Taylor, C., Patel, K., Jones, T., Kiely, F., De Stavola, B. L., and Sheer, D. (1993). *Brit. J. Cancer*, **67,** 128.
47. Kolluri, R. V., Manuelidis, L., Cremer, T., Sait, S., Gezer, S., and Raza, A. (1990). *Am. J. Haematol.*, **33,** 117.
48. Bienz, N., Leyland, M. J., and Hulten, M., personal communication.
49. Pinkel, D., Landegent, J., Collins, C., Fuscoe, J., Segraves, R., Lucas, J., and Gray, J. (1988). *Proc. Natl Acad. Sci. U.S.A.,* **85,** 9138.
50. Cremer, T., Lichter, P., Bordon, J., Ward, D. C., and Manuelidis, L. (1988). *Hum. Genet.,* **80,** 235.
51. Natarajan, A. T., Vyas, R. C., Darroudi, F., and Vermeulen, S. (1992). *Int. J. Radiat. Biol.,* **61,** 199.
52. Hulten, M. A., Gould, C. P., Goldman, A. S. H., and Waters, J. J. (1991). *J. Med. Genet.,* **28,** 577.
53. Rosenberg, C., Blakemore, K. J., Kearns, W. G., Giraldez, R. A., Escallon, C. S., Pearson, P. L., and Stetten, G. (1992). *Am. J. Hum. Genet.,* **50,** 700.
54. Dorin, J. R., Emslie, E., Hanratty, D., Farrall, M., Gosden, J., and Porteous, D. J. (1992). *Hum. Mol. Genet.,* **1,** 53.
55. Hutchison, N. and Weintraub, H. (1985). *Cell,* **43,** 471.
56. Griffin, D. K., Wilton, L. J., Handyside, A. H., Winston, R. M. L., and Delhanty, J. D. A. (1992). *Hum. Genet.,* **89,** 18.
57. Knights, P., Goldman, A. S. H., Kitts, J., and Hulten, M., personal communication.
58. Selig, S., Okumura, K., Ward, D. C., and Cedar, H. (1992). *EMBO J.,* **11,** 1217.
59. Goldman, A. S. H. and Hulten, M. A. (1992). *J. Med. Genet.,* **29,** 98.

5

Fine mapping of genes: the characterization of the transcriptional unit

M. ANTONIOU, E. DEBOER, and F. GROSVELD

1. Introduction

The fine mapping of genes includes a number of different techniques at several different levels of analysis. To include all these steps, to progress from a (partial) protein sequence or some detection method of the gene in question, to a complete determination of the transcriptional unit of that gene is beyond the scope of this chapter. The first part will, therefore, briefly indicate the various methods used to obtain cDNA and genomic DNA clones from a particular gene and will refer to a number of reviews on these topics. The second part will describe in more detail how to use such cDNA and/or genomic clones to analyse the transcriptional unit of the gene. The third part will indicate how to use this information to start the analysis of potential control regions of the gene.

2. cDNA and genomic DNA clones

The general principles to obtain cDNA or genomic DNA recombinants of a particular gene are indicated in *Figure 1* (1–3). The starting point is usually a (partially) purified protein and/or two different populations of RNA, one that expresses the gene of interest and another that does not. There are of course a number of exceptions, for example, in cases where complementation selection (4) or a biological assay (5) can be used, or where the final gene product is an RNA molecule such as ribosomal RNA. If the purified protein product can be obtained in nanomole quantities, it can be used to generate amino acid sequence data, which in turn allows the generation of (mixed) oligonucleotide probes that correspond to particular parts of the amino acid sequence (6). One or more of these probes can subsequently be used to screen a cDNA or genomic library to isolate the gene. These mixed oligonucleotide probes may also be used to amplify the desired sequence by polymerase chain reaction

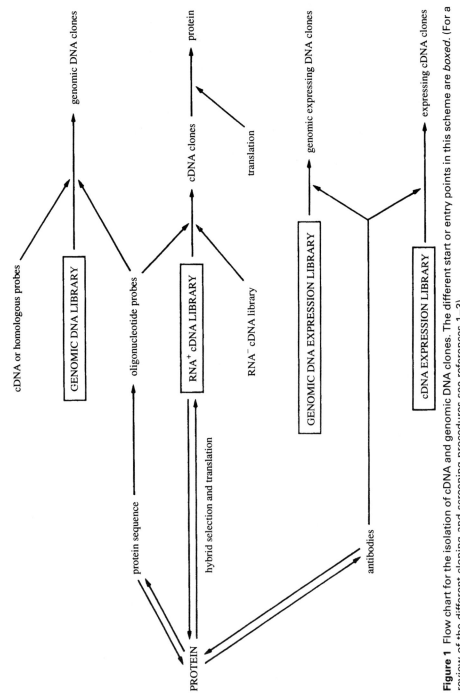

Figure 1 Flow chart for the isolation of cDNA and genomic DNA clones. The different start or entry points in this scheme are *boxed*. (For a review of the different cloning and screening procedures see references 1–3).

(PCR) of cDNA synthesized from expressing cells (59). Alternatively, the (partially) purified protein can be used to generate specific (monoclonal) anti-bodies against the protein. Such antibodies can be used to screen expression libraries, e.g. λgt11 cDNA libraries in *Escherichia coli* (1, 2), or eukaryotic cell expression libraries (7, 8). Once a cDNA or genomic clone is isolated, it can be used to obtain its genomic or cDNA counterpart by hybridization.

If the protein is not available, a differential screening method is employed between cDNA libraries from expressing and non-expressing tissues (9). The latter method is particularly useful to obtain a set of marker genes for particular tissues or developmental stages, rather than the isolation of indi-vidual genes with a known function or protein product (10).

3. Mapping the transcriptional unit

If we assume for illustrative purposes that we have either complete or partial genomic and cDNA clones for a gene containing three exons and two introns, we would have to analyse the following RNA polymerase II transcriptional unit; a primary transcript that is initiated at the first nucleotide of the first exon and terminated somewhere past the last nucleotide of the last exon. The primary transcript is processed to produce the mature mRNA at several positions; the introns are removed by a splicing process (11), and the 3′ end is formed by an endonucleolytic cleavage and polyadenylation (12). There are of course many exceptions to this general scheme, e.g. histone genes without introns or a poly(A) tail. A number of different techniques can be used to analyse the transcriptional unit:

- northern blots
- cDNA genomic DNA comparison
- R looping
- nuclease protection experiments
- 'transcription run-off' analysis

Each of these techniques have their particular advantages and limitations.

3.1 Restriction map

The simplest method which also provides a basis for all further mapping of the transcriptional unit, is construction of a restriction map using a number of different restriction enzymes (13). If both genomic and cDNA clones are available, a comparison of the two will give an immediate approximate indication of the localization of exons and introns. This information is roughly confirmed by 'Southern blot' (14) analyses of genomic DNA with cDNA (see *Protocol 1*). This type of analysis originally led to the first proposal by Jeffreys and Flavell (15) that the β-globin gene contained at least one intron between

Fine mapping of genes

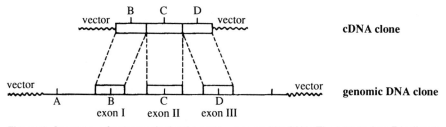

Figure 2 Structure of a theoretical eukaryotic gene and its DNA. The letters A to E indicate restriction enzyme sites.

two particular restriction sites. For our purpose, let us assume a gene, as shown in *Figure 2*. A genomic clone containing three exons and two introns, plus the restriction sites A to E, and a cDNA clone containing the sites B, C, and D.

Protocol 1. Southern blots

Solutions

- 20 × standard saline citrate (SSC)
  3 M NaCl
  0.3 M trisodium citrate pH 7.0

- 100 × Denhardt's solution
  1% bovine serum albumin
  1% Ficoll
  1% polyvinylpyrrolidone

- pre-hybridization solution
  3 × SSC
  0.1% sodium dodecyl sulphate (SDS)
  10 × Denhardt's solution
  50 µg/ml sheared and denatured salmon sperm DNA

- hybridization solution—as pre-hybridization mixture but including 10% dextran sulphate

Procedure

1. After electrophoresis in 90 mM Tris, 90 mM boric acid, 2.5 mM EDTA (pH 8.3), soak the 1.0 cm thick agarose gel (20 × 20 cm) for 40 min in 0.2 M HCl at room temperature. This will produce depurinated sites in the DNA which can be cleaved by alkaline hydrolysis.

2. Soak the gel for 40 min in 0.5 M NaOH, 1.5 M NaCl. If orange G is used as the indicator dye in the electrophoresis, it will turn red.

3. Soak the gel for 40 min in 1 M Tris (pH 8.0), 3 M NaCl, transfer the gel to a blotting set-up (51), and cover with a nitrocellulose filter and paper towels. If a nylon transfer membrane is used in place of nitrocellulose, the preceding neutralization step may be omitted. It is also not necessary to include a transfer buffer reservoir as part of the blotting set-up.

Blot for 3–5 h at room temperature which is sufficient to transfer 90% or more of the nucleic acid. If acid hydrolysis is omitted, the blotting should be much longer (overnight) with the aid of a transfer buffer reservoir, because high molecular weight DNA diffuses very slowly out of the gel. Blotting acid-treated gels overnight leads to a loss of DNA bound to the filter.

4. After blotting, wash the transfer membrane in 2 × SSC and fix the DNA to the filter by either baking for 2 h at 75–80°C, or by UV cross-linking (Stratalinker 2400, Stratagene).

5. Pre-hybridize the filter for a minimum of 15 min in pre-hybridization solution at 65°C.

6. Hybridize in 5–20 ml of hybridization solution containing denatured labelled probe for 6–12 h at 65°C. Depending on the probes, various competitors can be added to the hybridization, in addition to 50 μg/ml denatured salmon sperm DNA.

7. Wash the filters twice for 20 min in 3 × SSC, 0.1% SDS at 65°C, twice in 0.3 × SSC, 0.1% SDS at 65°C, and for stringent hybridization conditions once in 0.1 × SSC, 0.1% SDS at 65°C. After a final wash in 2 × SSC at room temperature for 2 min, dry the filters and expose them to X-ray film using cassettes and intensifying screens.

Note: there are several non-nitrocellulose filters on the market, which are blotted and hybridized as specified by the manufacturers. These filters are stronger, can be re-used more often, and at least some bind very low molecular weight fragments much more efficiency. Unfortunately, they are more expensive.

3.2 Northern blots

A second invaluable method used in transcription mapping is 'Northern' blotting (16). In this method RNA is electrophoresed in denaturing gels, transferred to nitrocellulose (or other) filters, and hybridized to cloned cDNA or genomic DNA to determine the size of the mature mRNA and its precursors (see *Protocol 2*). For the gene illustrated in *Figure 2* the smallest (and usually most intensely) hybridizing band would be the mature polyadenylated mRNA containing exons I, II, and III, (i.e. the same as the cDNA). In addition, higher molecular weight RNAs could be detected which correspond to precursor RNAs, in our example exons I, II, and III plus one or both of the introns. By using different probes, each of the exons and introns could be assigned in order (17). This method is very sensitive and can even be used to detect an intron in free linear and lariat form introns (18).

The combination of restriction maps and the use of several probes in Northern blots usually gives a fairly good global picture of the transcriptional unit, although any precise co-ordinates will not, as yet, have been established.

Protocol 2. Northern blots

Solutions

- 5 × running buffer
  200 mM 3-N-morpholino-propane sulphonic acid (MOPS), pH 7.0
  50 mM sodium acetate
  5 mM EDTA
- running buffer formaldehyde for northern blots
  40 mM 3-N-morpholino-propane sulphonic acid (MOPS, pH 7.0)
  10 mM sodium acetate
  1 mM EDTA
  2.2 M formaldehyde
- Sterile loading buffer
  50% glycerol
  1 mM EDTA
  0.4% bromophenol blue
  0.4% xylene cyanol
  or 20% Ficoll
  1 mM EDTA
  0.4% orange G

Procedure

1. Dissolve the RNA (1–20 μg) in 10–20 μl of running buffer formaldehyde and incubate at 55 °C for 15 min. Add 2–5 μl of sterile loading buffer.
2. Melt agarose (1–2% final concentration), cool to 60 °C, make 1 × with running buffer, and add formaldehyde, to 2.2 M.
3. After electrophoresis the marker lanes can be cut off, stained in 0.5 μg/ml ethidium bromide, and visualized on a U.V. transilluminator (51). Rinse the remainder of the gel in water for 5 min, and if necessary soak it for 40 min in 50 mM NaOH, 10 mM NaCl to partially hydrolyse the RNA which improves the transfer of high molecular weight material. Neutralize the gel in 1.0 M Tris (pH 7.0), 1.5 M NaCl, for 40 min, and transfer as Southern gels (*Protocol 1*).

3.3 'Run-on' transcription

The reverse procedure of the northern blot described above is a Southern blot using labelled RNA. This procedure is particularly important to localize the termination of transcription which proceeds past the poly(A) addition site at the 3' end of the mRNA (19). This technique is based on *in vitro* transcription in isolated nuclei in the presence of radiolabelled ribonucleotides. Under these circumstances, very little new initiation or RNA synthesis takes place. Only RNA chains initiated *in vivo* are elongated by several hundred nucleotides (20), and only partial processing of completed chains occurs (21).

The labelled RNA is isolated and hybridized to Southern blotted or dot-blotted restriction fragments of the cloned gene. The DNA fragments that are transcribed will hybridize and be visible on the autoradiograph. In the case of the mouse β-globin gene (*Figure 3*), the hybridization shows no signal for fragments upstream of the gene, an equal signal for fragments covering and directly downstream of the gene, and a decreasing signal in further downstream fragments. From these results the authors concluded that transcription termination does not take place at a discrete site, but over an area of about one kilobase (22). Because this method is not sensitive enough to detect very short hybrids and because the length of the hybrid is not measured, the boundaries of the transcription unit can only be determined to within a few hundred base pairs (23).

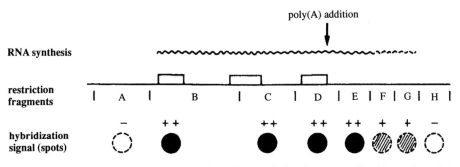

Figure 3 A theoretical 'run-on' hybridization. The restriction fragments A to H are spotted on to nitrocellulose filters and hybridized to 'run-on' labelled RNA ++, +, − indicate strong, weak, and zero hybridization signals, respectively.

Protocol 3. 'Run-on' transcription (52)

Solutions

- RSB
  10 mM Tris–HCl (pH 7.4)
  10 mM NaCl
  5 mM $MgCl_2$

Protocol 3. *Continued*

- freezing buffer for isolated nuclei
  50 mM Tris–HCl (pH 8.3)
  40% w/v glycerol
  5 mM $MgCl_2$
  0.1 mM EDTA

- 5 × 'run-on' transcription buffer
  25 mM Tris–HCl (pH 8.0)
  12.5 mM $MgCl_2$
  750 mM KCl
  1.25 mM ATP, GTP, and CTP

- 1 × SET
  1% SDS
  5 mM EDTA
  10 mM Tris–HCl (pH 7.4)

- TE
  10 mM Tris (pH 7.5)
  1 mM EDTA

- 20 × SSC (see *Protocol 1*)

- hybridization buffer (55)
  50% formamide (recrystallized or deionized)
  5% SSC (see *Protocol 1*)
  25 mM sodium phosphate
  0.1% sodium pyrophosphate
  5 × Denhardt's solution (see *Protocol 1*)
  1% SDS
  100 μg/ml sheared and denatured salmon sperm DNA
  200 μg/ml *E. coli* RNA

Procedure

1. Wash tissue culture cells with ice-cold phosphate-buffered saline (PBS) and, if necessary, trysinize from the plates. Wash complete tissues in PBS and homogenize with the aid of a Dounce homogenizer. All further steps are at 0°C unless stated otherwise.

2. Pellet the cells at 500 *g* for 5 min, wash them once in RSB, and repellet at 500 *g* for 5 min.

3. Resuspend the cells in RSB plus 0.5% NP-40 and release the nuclei by *gentle* agitation in a Dounce homogenizer.

4. Pellet the nuclei at 500 *g* for 5 min, resuspend them in 100 μl of freezing buffer at a concentration of 10^7–10^8 nuclei/100 μl, and freeze at −70°C.

5. Add 210 µl of thawed nuclei to 60 µl of 5 × 'run-on' buffer per reaction. Add 30 µl of [α^{32}P]UTP (3200 Ci/mmol) and incubate the suspension at 30°C for 30 min.

6. Add 15 µl of 5 µg/ml DNase I in 10 mM $CaCl_2$ and incubate at 30°C for 5 min. Stop the reaction by the addition of 5 × SET to a final concentration of 1 × SET.

7. Add proteinase K (20 mg/ml) to a final concentration of 200 µg/ml, and incubate for 45 min at 37°C.

8. Extract the RNA with an equal volume of phenol:chloroform (1:1) and centrifuge. Re-extract the interphase with 100 µl of 1 × SET and re-centrifuge. Make the combined aqueous phases 2.3 M NH_4OAc (10 M stock), and add an equal volume of isopropanol. After 15 min at -70°C pellet the nucleic acid in a microfuge for 10 min.

9. Dissolve the pellet in 100 µl of TE. Run the sample through a G-50 spin column to remove unincorporated labelled nucleotides. Make the eluate 0.2 M NaOH and keep it on ice for 10 min. Add Hepes (from a 1 M stock solution) to a final concentration of 0.24 M for neutralization, and precipitate the nucleic acid by the addition of 2.5 volumes of 100% ethanol, stored at -20°C. Pellet the nucleic acid in a microcentrifuge (10 min). Resuspend the pelleted labelled RNA in 2–5 ml of hybridization buffer.

10. Bind the restriction fragments (0.5–5 µg) to the nitrocellulose filters with the aid of a Schleicher and Schull Slot-Blot apparatus as suggested by the manufacturers.

11. Pre-hybridize the filters at 42°C in hybridization buffer for 2 h and transfer to 1–2 ml of hybridization solution containing the [^{32}P]RNA. Hybridize for 36 h at 42°C. Wash the filters twice for 15 min in 2 × SSC, 0.1% SDS at room temperature, and once at 60°C in 0.1 × SSC, 0.1% SDS for 30 min.

12. Incubate the filters for 30 min at 37°C in 2 × SSC containing 10 µg/ml RNase A, and wash again in 2 × SSC at 37°C for 2 h.

13. Expose the filters to Kodak XAR film in cassettes with intensifying screens at -70°C. Quantitate the 'slots' by scanning or counting in a scintillation counter. Correct the final quantities for the length of the DNA fragment and (if known) the T content (the label was only in UTP).

3.4 R looping

A more visual localization of intron–exon junctions is possible using electron microscopy (EM) and R looping (24). Because of the techniques and machinery involved, this technique will not be readily available everywhere and requires considerable EM experience. We will, therefore, not provide a laboratory protocol, but only refer to published work. It is, however, very informative and is definitely very useful when a gene with a large number of exons and introns has to be handled. Moreover, it will demonstrate whether within the limits of detection all the mRNA is accounted for in terms of cloned DNA, or whether the cloned gene is still incomplete. The method is based on the observation that DNA–RNA hybrids are of the A-type helix, whereas duplex DNA is in the B-type form. This difference in structure causes a different stability in different solvents. In formamide DNA–RNA hybrids are more stable, while in aqueous solutions duplex DNA is more stable (25, 26). In order to form R loops, duplex DNA and RNA are hybridized in a formamide solution close to the melting temperature of the DNA duplex. Transient single stranded regions will hybridize to the complementary RNA and form stable hybrids, displacing the complementary DNA strand which is visible as a single strand (R) loop (*Figure 4A*). By using different restriction enzymes it is possible to position the transcriptional unit within about 100 base pairs. In our example we would expect the type of structure shown in *Figure 4A*, which was first observed for the globin and ovalbumin genes (17, 27). Structures such as shown in *Figure 4B* would indicate that the DNA clones are incomplete. Alternatively, when the DNA clones are deliberately shortened, (e.g. by cleavage at site B) the combination of *Figure 4A* and *4B* would map the transcript.

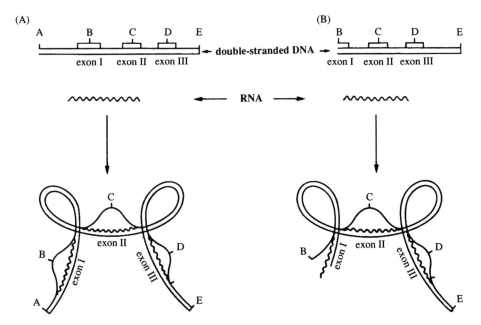

Figure 4 'R-looping'. The double stranded genomic DNA (A to E or B to E) is hybridized to RNA under R looping conditions.

3.5 Nuclease S1 protection and Southern blots

The remainder of the techniques which we will discuss are all based on the same very basic hybridization techniques we have already mentioned, but are much more precise in their measurement and use of enzymatic reactions. The methods are mostly based on nuclease S1 protection experiments that were originally developed by Berk and Sharp (28) to map the transcripts of early adenovirus genes. Nuclease S1 from *Aspergillus oryzae* is an enzyme that degrades single stranded DNA or RNA (29). Double stranded DNA, RNA, and DNA–RNA hybrids are resistant to degradation unless very large amounts of the enzyme are used. A similar enzyme is mung bean nuclease (30), although this enzyme will not cleave the DNA strand opposite a nick in a duplex, while nuclease S1 will. These enzymes can, therefore, be used to measure the size of a DNA–RNA hybrid by cleavage of the non-hybridized single stranded nucleic acid. If we apply the technique to our example, we could expect the following (*Figure 5*). The RNA preparation which contains

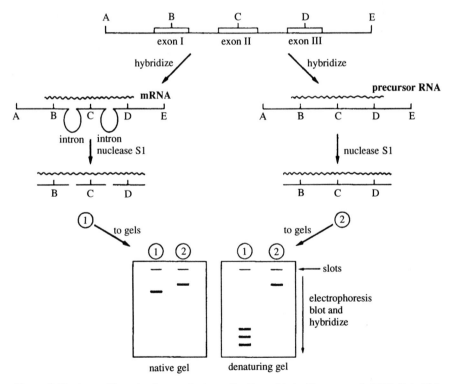

Figure 5 Nuclease S1 protection analysis on Southern blots. The genomic DNA (A to E) is hybridized to an RNA population containing processed mRNA and precursor RNA. The RNA–DNA hybrids are treated with nuclease S1 [(1) and (2)] and analysed on agarose gels.

the primary transcript, processing intermediates, and the mature mRNA is hybridized with a denatured genomic DNA fragment under high formamide conditions where DNA–DNA annealing is prevented. Several different hybrids will be formed. The primary transcript will hybridize to the DNA over its full-length, whereas in the case of spliced mRNA or intermediates, hybrid molecules will be formed that contain loops of intron DNA. (The opposite would be obtained with cDNA, i.e. the precursor would be contained in a hybrid with RNA loops, while the mRNA would be present in a co-linear hybrid.) The hybrids are then treated with nuclease S1, which destroys the single stranded nucleic acid but leaves the hybrids intact. Consequently, the primary transcript will be reduced to a full-length perfect duplex, by removal of the excess single stranded DNA and RNA tails. The processed transcript will similarly be reduced to a hybrid with two nicks in the DNA at the positions where the intron DNA loops were removed by the nuclease S1. The two types of molecules can be distinguished by agarose gel electrophoresis under denaturing and non-denaturing conditions. The RNA will be destroyed by alkali in the denaturing gel and will still be present in the hybrid on the non-denaturing ('native') gel. The primary transcript will be the same full-length fragment on both gels. The spliced transcript would become visible as a single band on the native gel and as three separated bands, each the size of the individual exons, on the alkaline gel. Usually, unlabelled DNA fragments are used in the hybridization to RNA, and nuclease S1 protected fragments are only detected after Southern blotting of the gel and hybridization to labelled DNA probes from the same gene. This method has the advantage (over subsequent methods) that it can be used with long DNA fragments containing multiple exons and introns. In one experiment a fairly good picture of the transcriptional unit can be obtained. This is especially true when two-dimensional electrophoresis is used, one dimension at neutral pH, the second dimension at alkaline pH, and when the blots are hybridized to different probes corresponding to different parts of the gene (31). The disadvantage (particularly for long fragments on agarose gels), is that the sizes are not absolutely precise, due to inaccuracies in the marker sizing and diffusion of the bands during the blotting procedure.

3.6 Nuclease S1 protection of end-labelled probes

Once the (approximate) position of the exon, or later when some restriction sites in an exon are known, end-labelled probes can be used for a very precise localization of exon and intron borders. The procedure is a variation of the Berk and Sharp method, but instead of using unlabelled probes, restriction fragments are used that are only labelled at the ends (32). For example, for our test gene we could use DNA probes from restriction sites A to B and B to C. The fragments can be labelled in two ways, either at the 5′ ends using polynucleotide kinase (33), or at the 3′ ends by 'filling-in' or 'replacement' synthesis using DNA polymerase I (Klenow fragment) (34), T4 DNA polymerase (35), or reverse transcriptase (36). The labelled DNA fragment is subsequently denatured and hybridized to RNA in high formamide concentrations. Higher sensitivity (almost always unnecessary) can be obtained by separating individual strands of the DNA probe by polyacrylamide gel electrophoresis (37) prior to use. After hybridization, the samples are treated with nuclease S1 which will degrade all the single stranded regions. The protected hybrids are subsequently denatured and electrophoresed on polyacrylamide gels in the presence of 6 M urea, using a sequence ladder of the same fragment as a very precise size marker (*Figure 6*). In our example (*Figure 6*) the kinased AB fragment would measure the 5′ end initiation point of the mRNA. The kinased BC fragment would measure the splice acceptor site (fragment 3) of the first intron, whereas the 'filled-in' fragment would measure the splice donor site of the first intron (fragment 4). Fragments 2 and 5 would not show up in the analysis because they are not labelled. *Figure 7* is

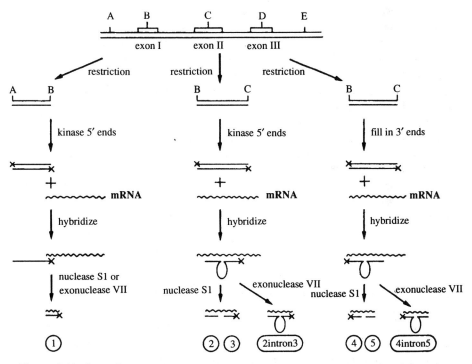

Figure 6 Nuclease S1 protection analysis with end-labelled DNA probes. The genomic DNA fragments AD and BC are labelled at the 5′ or 3′ ends by 'kinase' or 'fill-in' reaction and hybridized to mRNA. The RNA–DNA hybrids are treated with nuclease S1 or exonuclease VII.

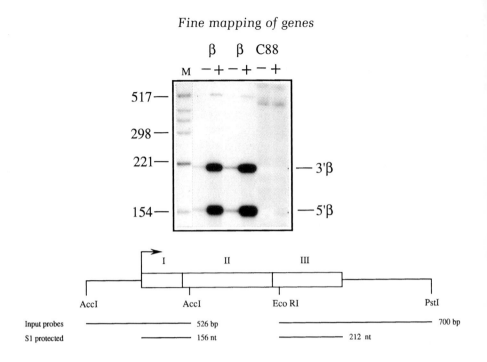

Figure 7 Nuclease S1 protection analysis of human β-globin gene transcripts. Probes were prepared from a β-globin gene from which the introns had been removed as illustrated *above*. A 5′ 526 bp *Acc*I fragment labelled by kinasing, gives a 156 nucleotide(nt) S1 protected product(5′β). A 3′ 700 bp *Eco*RI–*Pst*I fragment labelled by 'filling-in' with reverse trascriptase, gives a 212 nt S1 protected product(3′β). Both probes were labelled with [32]P. These probes respectively map the 5′ cap site and poly(A)-addition site of the transcript. RNA was from murine erythroleukaemia cells stably transfected with the human β-globin gene (β) whose expression is driven by the locus control region (see 56). Cells were taken both before(−) and after (+) induced erythroid differentiation. C88 refers to RNA from untransfected (negative control) cells. The size markers (M), are a *Hinf*I digest of the plasmid pBR322.

an example of a 5′ and 3′ end analysis of mRNA transcribed off the human β-globin gene. The 5′ end probe in the example is represented by a kinased *Acc*I restriction fragment. For the 3′ end analysis a 'filled-in' *Eco*RI– *Pst*I fragment was used. A G + A sequence ladder would provide a very precise measurement of the mRNA 5′ end. It is obvious that this is a very precise method because it provides a border at an exact distance from a known restriction site. The disadvantage is that multiple probes have to be used to map each border of all the exons; in our example, at least four fragments for six S1 assays (AB, BC 2 ×, CD 2 ×, and DE). Moreover, it should be remembered that this technique has to be supplemented with at least one other technique such as primer extension when the number of exons is unknown (see below at primer extension).

Protocol 4. Nuclease S1 protection assay

Solutions

- hybridization buffer
  80% formamide (recrystallized or deionized)
  40 mM Pipes (pH 6.7)
  0.4 M NaCl
  1 mM EDTA
  store at −20°C

- 10 × S1 digestion buffer
  300 mM sodium acetate (pH 4.5)
  2.8 M NaCl
  45 mM zinc acetate
  store at 4°C

- loading buffer
  7 M urea
  5 mM Tris-borate (pH 8.3)
  1 mM EDTA
  0.1% xylene cyanol
  0.1% bromophenol blue
  store at −20°C

Procedure

1. Label 100 ng of probe by kinasing or 'filling-in', which is enough for 20 lanes approximately. Otherwise, go directly to step **2**.

2. Dissolve the probe in hybridization buffer.

3. Precipitate the RNA (1–50 μg of RNA per lane depending on how abundant the mRNA is) with 100% ethanol (2.5 volumes) in the presence of 0.3 M sodium acetate pH 5.5, spin, and dissolve the pellet in 10–30 μl of hybridization buffer with the probe.

4. Place at 90°C for 5 min to denature the RNA and DNA probe, transfer immediately into a 52°C water-bath, and hybridize for a minimum of 6 h to overnight. Have the two water-baths next to each other. We do the hybridization in Eppendorf tubes and keep them sealed by clamping the lids between the holding rack and a solid cover (plastic or metal). Keep the surface of the water in the two baths to the neck of the tubes.

5. While the tubes are still in the water-bath, just open the lids; do not lift out until you add 10 volumes of 1 × digestion buffer containing 100 units of nuclease S1 (Boehringer). Then vortex and put in ice until all the tubes are done.

Protocol 4. *Continued*

6. Transfer into a 20°C water-bath for 2 h (certain probes need different temperatures, 20–40°C), extract once with an equal volume of phenol: chloroform (1:1 v/v), and precipitate with 2.5 volumes of 100% ethanol at −20°C for 30–60 min in the presence of 10 μg *E. coli* tRNA as carrier if less than 10 μg RNA was used in the hybridization. Dissolve the pellet in 5 μl of S1 loading buffer.

7. Denature at 90°C for 5 min and run on a sequence gel (37) or agarose gel if blotting is required.

3.7 Exonuclease VII

A very useful complementation of nuclease S1 analysis is the use of *E. coli* exonuclease VII. This enzyme is a progressive exonuclease that cleaves small oligonucleotides from the 5′ and 3′ ends of single stranded DNA (38). It can therefore be used in a similar fashion to nuclease S1 or mung bean nuclease to map a transcriptional unit (39), with the exception that it will not cleave any single stranded loops flanked by double stranded nucleic acid. For example, the hybrids shown in *Figure 6* would yield a different product with exonuclease VII. Fragment 1 would still be found, but fragments 2, 3, 4, and 5 would now be detected as larger fragments, i.e. 2/3 and 4/5 are still connected by an intron. Taken together with the nuclease S1 analysis, this would provide immediate information not only about the sizes of the exons, but also the introns. The second useful application of exonuclease VII is the detection of nuclease S1 artefacts created by partial melting of very A + T rich regions ('breathing') or repetitive sequences ('slipping') in a DNA–RNA hybrid.

Protocol 5. Exonuclease VII digestion

Solutions

- exonuclease VII buffer
  30 mM KCl
  10 mM Tris (pH 7.8)
  10 mM EDTA
- other solutions as described in *Protocol 4*

Procedure

1. The hybridizations are carried out exactly as described for the nuclease S1 protection assay (*Protocol 4*). Proceed as from step **4**.

2. Add 500 μl of exonuclease VII buffer with 4 U/ml exonuclease VII. Incubate for 2 h at 37°C. Extract with phenol:chloroform, ethanol precipitate, and proceed exactly as described for the nuclease S1 protocol steps **6** and **7**.

3.8 T7, T3, or SP6 polymerase probes

T7, T3, and SP6 RNA polymerase are DNA dependent RNA polymerases found in cells infected with bacteriophage T7 (41), T3 (40), or SP6 (42). Each of these polymerases initiates RNA synthesis from T7, T3, or SP6 promoter sequences with very high specificity. This property can be used to synthesize single stranded RNA from any sequence linked to such a promoter. This system was initially developed to produce large amounts of labelled precursor to study RNA processing (43). However, the RNA can also be used more effectively than DNA as a probe for DNA and RNA, because DNA–RNA and RNA–RNA hybrids in particular, are more stable than DNA–DNA hybrids. RNA–RNA hybrids have a normal melting temperature (T_m) of 94–100°C or 75–85°C in 50% formamide, whereas DNA–DNA hybrids usually have a T_m of 85–95°C in 50% formamide (44). The uniformly labelled RNA probes are very sensitive and allow the detection of as little as one picogram of mRNA on Northern blots (45). This is largely due to the single stranded nature of the probe which results in no competition with complementary sequences such as seen with double stranded DNA probes. A practical disadvantage of this system is the fact that the DNA sequence of interest first has to be cloned next to the T7, T3, or SP6 promoter. Since it is quicker to isolate a restriction fragment than to clone it, this procedure usually means a loss of time for a single series of experiments. However, a definite gain of time is obtained when the same part of the transcriptional unit is repeatedly analysed (one cloning is, in principle (see below), quicker and cheaper than repeated fragment isolation). There are a number of vectors commercially available which contain T7, T3, or SP6 promoters. Usually, the DNA sequences are cloned into a polylinker attached to the promoter, e.g. fragment AC of our theoretical gene (*Figure 8*). The recombinant plasmid is linearized at the downstream restriction site (C in *Figure 8*) of the inserted fragment. In the presence of labelled ribonucleotides, the polymerase initiates synthesis at the specific promoter and produces a labelled RNA transcript up to the end of the insert where the enzyme runs off the linear template. Each polymerase can transcribe the template several times and up to ten micrograms of RNA can be produced from one microgram of template DNA. The uniformly labelled RNA probe is then used in a similar fashion to the nuclease S1 DNA probes, with the exception that T1 and pancreatic ribonuclease (RNase A) are used in the protection experiment instead of nuclease S1. It should be pointed out that the synthesis is not always complete and as a consequence, the full-length synthesized probe is often first purified by acrylamide gel electrophoresis to obtain optimal results. Unfortunately, this step eliminates one of the advantages (convenience) of this system. In our example the AC probe would protect the complete first exon and half of the second exon when hybridized to mRNA. Of course, if the transcriptional unit is unknown, this method has the same advantage as S1 blotting or Northern blotting techniques when compared to

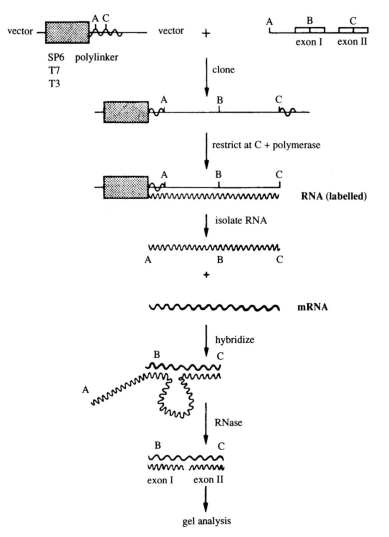

Figure 8 Labelled RNA probes. The genomic DNA fragment (AC) is cloned in a polylinker between sites A and C. The recombinant DNA is cleaved at C and trascribed into labelled RNA. After hybridization to the mRNA, the RNA–RNA hybrid is cleaved with RNase and analysed on denaturing polyacrylamide gels (37).

end-labelled probes, that is, it gives precise sizes but not the co-ordinates of the transcript in relation to a known position. To obtain precise co-ordinates, at least one other analysis with a different probe that has been restricted at a different site has to be performed. Lastly, it should be pointed out that very similar (but more cumbersome) procedures are available to synthesize uniformly labelled DNA probes from a single stranded DNA template, using the Klenow fragment of DNA polymerase I and DNA primer (46, 47).

Protocol 6. Single stranded RNA probes

Solutions

- 5 × SP6 buffer
  200 mM Tris–HCl (pH 7.5)
  30 mM MgCl$_2$
  10 mM spermidine

Note: the manufacturers of SP6, T7, and T3 polymerases now provide the appropriate reaction buffer with the enzyme, which can be used in place of the above.

- 10 × MS buffer
  100 mM Tris–HCl (pH 7.5)
  100 mM MgCl$_2$
  0.5 M NaCl

- solution hybridization buffer—as in *Protocol 4*

- RNase buffer
  0.3 M NaCl
  10 mM Tris–HCl (pH 7.5)
  5 mM EDTA

- 10% sodium dodecyl sulphate

- gel loading buffer—as in *Protocol 4*

- blot hybridization buffer
  50% formamide
  50 mM NaPO$_4$ (pH 6.5)
  5 × SSC
  0.1% SDS
  5 × Denhardt's solution
  200 μg/ml denatured salmon sperm DNA

Procedure
Transcription (D. Ish-Horowitz, personal communication)–hybridization (53). The recipe for a radiolabelled probe is given below. For the synthesis of probes conjugated with non-radioactive adjuncts, (e.g. biotin or digoxygenin), we follow the manufacturer's guidelines.

1. Incubate the ^{35}S-probe mixture for ≥ 1 h at 30 °C (up to 4 h).

		Final concentration
5 × SP6 buffer	4 μl	1 ×
BSA 5 mg/ml	0.4 μl	100 μg/ml
Triton X-100 5%	0.5 μl	0.1%

Protocol 6. *Continued*

		Final concentration
rNTP (C + A + G 10 mM each)	1 µg	0.5 mM
DTT 0.5 M	0.5 µl	12 mM
SP6 polymerase (dil 1:10)	0.5 µl (= 1.25 U)	
RNasin (inhibitor, 1 U/µl)	0.5 µl	25 U/ml
[^{32}P]- or [^{35}S]UTP	10 µl	~4 µM
1.2 µg linearized DNA (2 kb insert)		~60 µg/ml
add H$_2$O to a final volume of 20 µl		

2. Make reaction 1 × MS, add 1/10 volume RNase-free DNase (54) (20–50 µg/ml), and incubate at 37°C for 30 min.

3. Add EDTA to 20 mM, SDS to 0.2%, 5 µg tRNA carrier, and spin through Biogel P-60 or Sephadex G-50 column.

4. Separate the full-length transcript on polyacrylamide gels (37) if the final analysis has a high background due to incomplete RNA synthesis or incomplete digestion of the DNA template.

Notes: the K_m for UTP is high, UTP concentration must be >2 µM or there will be substantial premature termination, >10 µM is better. If necessary, dilute the label with cold triphosphate. CTP has a lower K_m and is better for making full-length probes. The lower incubation temperature (30°C) and 0.1% Triton favours longer transcripts. Salt in the reaction promotes premature termination and should be avoided. Also, NH$_4^+$ is a potent inhibitor of the SP6 polymerase. For hotter probes, use T7 polymerase. T7 polymerase also has a lower K_m for UTP so it is easier to use with labelled UTP.

The above recipe works for T7 with the following modifications:

- use about four units of T7 polymerase per reaction
- avoid labelled GTP, this has a high K_m

5. Dissolve the test RNA (5–40 µg) and the labelled probe RNA in 30 µl of solution hybridization buffer. Denature at 85°C for 5 min and incubate (>8 h) at 45°C. Different temperatures are optimal depending on the G + C content of the hybrid.

6. Add 300 µl of RNase buffer, containing 40 µg/ml RNase A, and 2 µg/ml RNase T1, and incubate for 1 h at 39°C. Again different temperatures might be used to obtain the best signal-to-background ratio.

7. Stop the RNase digestion by the addition of 20 µl 10% SDS and 50 µg of proteinase K. Incubate for 15 min at 37°C. Extract with an equal volume of phenol:chloroform (1:1 v/v) and precipitate the labelled RNA hybrids with 5 µg tRNA in 2.5 volumes of 100% ethanol. Spin, dissolve the pellet in gel loading buffer, and run on 8 M urea-polyacrylamide gels (37).

8. If the probes are required for Southern or Northern blots, proceed from step **3** to dissolving the probe in blot hybridization buffer and hybridize at 55–60°C. Wash as for Southern blotting protocol.

3.9 Primer extensions

The primer extension method is most commonly used to detect the start site(s) and 5′ end of splice junctions of the transcriptional unit. Moreover, just like S1 or T1 protection analysis, it can be used routinely to quantitate the levels of a particular mRNA (48, see *Figure 10*). A labelled oligonucleotide or small restriction fragment is hybridized to the template mRNA and used as a primer for the synthesis of a complementary DNA by reverse transcriptase in the presence of unlabelled nucleotides. Alternatively, an unlabelled primer can be used and extended with radioactively labelled nucleotides, although the first method usually gives less background in the analyses. Either way, the net result is a labelled cDNA of defined length as synthesis stops at the 5′ end of the RNA (see *Figure 9*,

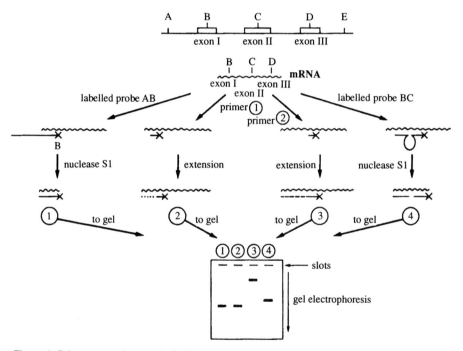

Figure 9 Primer extension analysis. The mRNA is analysed by primer extension using two labelled primers; primer 1 which is homologous to part of exon I and primer 2 which is homologous to part of exon II. Each primer is extended and the product analysed on a denaturing polyacrylamide gel (lanes 2 and 3). For comparisons, lanes 1 and 4 contain the products of a nuclease S1 protection experiment with labelled genomic DNA probes AB and BC respectively.

primer 1). Measurement of the cDNA on polyacrylamide gels gives a co-ordinate for the 5' end of the mRNA in relation to the known position of the primer. This method, together with a nuclease S1 protection assay is usually regarded as 'solid' evidence for the position of the 5' end of the mRNA, (e.g. 49). It is important to note that a number of independent but complementary techniques have to be used for definitive results to be obtained. Each method alone does not provide conclusive evidence, primer extension can give premature stops in the synthesis (strong stops), while nuclease S1 protection analysis can cleave at breathing and slippage positions (see above). Moreover, a particular nuclease S1 probe that is isolated from the first exon present in the cloned DNA might not necessarily correspond with the 5' end of the mRNA, if another unknown exon is further upstream. Such a situation would be readily detected by the combination of procedures because the primer extension and the nuclease S1 protection analysis would give different co-ordinates (*Figure 9*, primer 2). Lastly, the product of a primer extension can be directly sequenced if an end-labelled primer is used. This is very convenient to locate the exact site of initiation of the mRNA and the exact intron–exon borders when the sequence is compared to the genomic DNA. Such an analysis has, for example, been used to definitely characterize the aberrant splicing event that takes place in a particular β-globin thalassaemia (50).

Protocol 7. Primer extension

Solutions

- hybridization buffer
  400 mM NaCl
  10 mM Pipes (pH 6.4)
- 5 × primer extension buffer
  250 mM Tris–HCl (pH 8.2)
  50 mM DTT
  30 mM $MgCl_2$
- 10 mM dNTP's—mixture of four nucleotide triphosphates
- 10 mg/ml actinomycin D

Procedure

1. After labelling, isolate the synthetic primer or the primer restriction fragment from a preparative agarose gel or polyacrylamide gel (depending on its size) and ethanol precipitate.

2. Dissolve 10–30 μg of total RNA and about 0.2 pmol of the labelled primer in 10 μl of hybridization buffer. Seal the mixture in a glass capillary, denature at 90°C for 2 min, and hybridize 45 min to overnight at 30°C.

Note: the amounts of RNA and primer depend on the relative abundance in the total RNA population. The hybridization temperature depends on the length and sequence of the primer; the conditions described above are for a 30-mer. In general, annealing should be 5°C below the T_m.

3. Primer extension is started by delivering the hybridization mixture into the extension mixture containing 1 × primer extension buffer, 0.5 mM of each dNTP, 2.5 μg actinomycin D, and 10 U of reverse transcriptase. Incubate the reaction at 41°C for 1 h. Stop the reaction by the addition of 0.1% SDS, 10 mM EDTA (final concentrations), extract the products with phenol:chloroform, and precipitate with ethanol.

4. Run the extension products, with or without sequencing, on denaturing polyacrylamide gels (37).

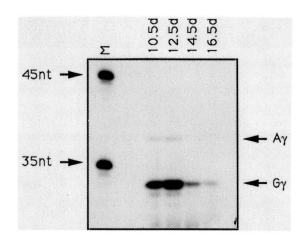

Figure 10 Quantitative analysis by primer extension of the developmental pattern of expression of the human γ-globin genes in transgenic mice. RNA was from embryonic yolk sac (10.5 days) and fetal liver (12.5 to 16.5 days). The assay takes advantage of a short region of non-homology of the 3' end of exon III in the otherwise completely homologous human γ-globin genes. By using ddTTP in the reaction, the Gγ-specific extended product will terminate just before the region of non-homology (32 nucleotides), whereas the Aγ-specific extended product will terminate just after this region (42 nucleotides).

4. Polymerase chain reaction (PCR)

Once a cDNA clone has been obtained and sequenced, then the polymerase chain reaction (PCR; 57) may be used to map the positions of any introns in the corresponding native gene. However, this procedure is only of practical value in cases where genomic DNA is available in limited amounts and which would thus preclude the use of the methods described earlier in this chapter.

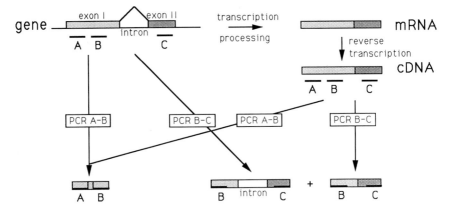

Figure 11 Analysis of intron/exon structure and quantitation of mRNA by PCR. A, B, and C refer to oligonucleotide primers complementary to exonic sequences.

The principle of this approach is outlined in *Figure 11*. It consists of taking pairs of oligonucleotide primers complementary to various positions throughout the cDNA. PCR is then performed with the cloned cDNA and genomic DNA. If the pair of primers used lie within the same exon, then the size of the PCR product will be the same with both samples (*Figure 11*; PCR A–B). Alternatively, if the primers lie in adjacent exons, then the size of the product from the genomic DNA will be larger than that from the cDNA by an amount corresponding to the length of the intervening intron (*Figure 11*; PCR B–C). PCR may also be used to quantitate and map mRNA. This again is useful for analysis of low abundance sequences and/or when tissue is only available in small quantities, (e.g. see 58, 59). Starting with total RNA, an initial cDNA synthesis from the corresponding mRNA is first performed using an appropriate complementary oligonucleotide primer and reverse transcriptase. A standard PCR is then carried out with a second set of primers to amplify the cDNA product. In order to obtain a quantitative comparison of the mRNA in different samples, it is important to limit the number of PCR cycles to the linear phase of the reaction (this will need to be determined empirically).

The principles and methods of PCR have been covered in great detail in two other volumes in this series to which we refer the interested reader (59, 60).

Acknowledgement

The excellent secretarial assistance of Cora O'Carrall is gratefully acknowledged.

References

1. Glover, D. M. (ed.) (1985). In *DNA Cloning: A Practical Approach*, Vol. 1. IRL Press, Oxford.
2. Karn, J. (1983). In *Techniques in the Life Sciences. Nucleic Acid Biochem. B501.* Vol. B5. (ed. R. A. Flavell). Elsevier Scientific Publishers Ireland Ltd.
3. Grosveld, F. G. and Dahl, H. H. M. (1983). In *Techniques in the Life Sciences: Nucleic Acid Biochem. B502.* Vol. B5. (ed. R. A. Flavell). Elsevier Scientific Publishers Ireland Ltd.
4. Chang, A., Nurnberg, J., Kaufman, R., Ehrlich, H., Schimke, R., and Cohen, S. (1978). *Nature,* **275,** 617.
5. Nagata, S., Taira, H., Hall, A., Johnsrud, L., Streuli, M., Escodi, J., Boll, W., Cantell, K., and Weissmann, C. (1980). *Nature,* **284,** 316.
6. Wallace, M. B., Johnson, M. J., Hirose, T., Miyake, T., Kawashima, E. K., and Itakura, K. (1982). *Nucleic Acids Res.,* **9,** 879.
7. Kuhn, L. C., McClelland, A., and Ruddle, F. H. (1984). *Cell,* **37,** 95.
8. Littman, D. R., Thomas, Y., Madon, P. J., Chess, L., and Axel, R. (1985). *Cell,* **40,** 237.
9. Kavathas, P., Sukhatine, V. P., Herzenberg, C. A., and Parnes, J. R. (1984). *Proc. Natl Acad. Sci. U.S.A.,* **81,** 7688.
10. Anderson, D. J. and Axel, R. (1985). *Cell,* **42,** 649.
11. Lamond, A. (1991). *Current Opinion in Cell Biol.,* **3,** 493.
12. Birnstiel, M. L., Busslinger, M., and Strub, K. (1985). *Cell,* **41,** 349.
13. Boseley, P. G. (1983). In *Techniques in the Life Sciences B511, Nucleic Acid Biochem.* Vol. B5. (ed. R. A. Flavell). Elsevier Scientific Publishers Ireland Ltd.
14. Southern, E. M. (1975). *J. Mol. Biol.,* **98,** 503.
15. Jeffreys, A. J. and Flavell, R. A. (1977). *Cell,* **12,** 1097.
16. Alwine, J. C., Kemp, D. J., and Stark, G. R. (1977). *Proc. Natl Acad. Sci. U.S.A.,* **74,** 5350.
17. Chambon, P., Benoist, C., Breathnach, R., Cochet, M., Cannon, F., Gerlinger, P., Knist, A., LeMeur, M., LePennec, J. P., Mandel, J. L., O'Hare, K., and Perrin, F. (1979). In *From Gene to Protein. Information transfer in normal and abnormal cells*, Vol. 16 (ed. R. Russell, K. Brew, H. Faber, and J. Schultz), pp. 55–78. Academic Press, New York.
18. Zeitlin, S. and Efstratiadis, A. (1984). *Cell,* **39,** 589.
19. Hofer, E. and Darnell, J. E. (1981). *Cell,* **23,** 585.
20. Weber, J., Jelinek, W., and Darnell, J. E. (1977). *Cell,* **10,** 611.
21. Blanchard, J. M., Weber, J., Jelinek, W., and Darnell, J. E. (1979). *Proc. Natl Acad. Sci. U.S.A.,* **75,** 5344.
22. Salditt-Georgief, M. and Darnell, J. E. (1983). *Proc. Natl Acad. Sci. U.S.A.,* **80,** 4694, and **81,** 2274.
23. LeMuer, M. A., Galliot, B., and Gerlinger, P. (1984). *EMBO J.,* **3,** 2779.
24. White, R. L. and Hogness, D. S. (1977). *Cell,* **10,** 177.
25. Birnstiel, M. L., Sells, B. H., and Purdom, T. (1972). *J. Mol. Biol.,* **63,** 21.
26. Casey, J. and Davidson, N. (1977). *Nucleic Acids Res.,* **4,** 1539.
27. Tilghman, S. M., Tiemeier, D. C., Seidman, J. G., Peterlin, B. M., Sullivan, M., Maizel, J. V., and Leder, P. (1978). *Proc. Natl Acad. Sci. U.S.A.,* **75,** 725.

28. Berk, A. J. and Sharp, P. A. (1977). *Cell*, **12**, 721.
29. Vogt, V. M. (1973). *Eur. J. Biochem.*, **33**, 192.
30. Laskowski, M. (1980). In *Methods in Enzymology*, Vol. 65 (ed. L. Grossman and K. Moldave), pp. 263–76. Academic Press, New York.
31. Grosveld, G. C., Koster, A., and Flavell, R. A. (1981). *Cell*, **23**, 573.
32. Weaver, R. F. and Weissmann, C. (1979). *Nucleic Acids Res.*, **6**, 1175.
33. Richardson, C. C. (1971). *Proc. Natl Acad. Sci. U.S.A.*, **2**, 815.
34. Jacobson, H., Klenow, H., and Overgaard-Hansen, K. (1974). *Eur. J. Biochem.*, **45**, 623.
35. Sanger, F. and Coulson, A. R. (1975). *J. Mol. Biol.*, **94**, 441.
36. Verma, I. M. (1977). *Biochim. Biophys. Acta*, **473**, 1.
37. Maxam, A. M. and Gilbert, W. (1980). In *Methods in Enzymology*, (ed. L. Grossmann and K. Moldave), pp. 499–560. Academic Press, New York.
38. Chase, J. W. and Richardson, C. C. (1964). *J. Biol. Chem.*, **249**, 4545.
39. Berk, A. J. and Sharp, P. A. (1978). *Cell*, **14**, 695.
40. Chakrabarty, P. R., Sarkar, P., Huang, H. H., and Maitra, U. (1973). *J. Biol. Chem.*, **248**, 6637.
41. Chamberlain, M., McGrath, J., and Waskell, L. (1970). *Nature*, **228**, 227.
42. Butler, E. T. and Chamberlain, M. J. (1982). *J. Biol. Chem.*, **257**, 5772.
43. Green, M. R., Maniatis, T., and Melton, D. A. (1983). *Cell*, **32**, 681.
44. Cox, K. H., deLeon, D. V., Augerer, C. M., and Augerer, R. C. (1984). *Dev. Biol.*, **101**, 485.
45. Zinn, K., Maio, D., and Maniatis, T. (1983). *Cell*, **34**, 865.
46. Rica, G. A., Taylor, J. M., and Kalinyak, J. E. (1982). *Proc. Natl Acad. Sci. U.S.A.*, **79**, 724.
47. Antoniou, M., Guzman, K., Chakraborty, S., and Banerjee, M. R. (1985). *J. Biophys. Biochem. Meth.*, **11**, 208.
48. McKnight, S. L., Garis, E. R., and Kingsbury, R. (1981). *Cell*, **25**, 385.
49. Giguere, V., Ishobe, K.-I., and Grosveld, F. G. (1985). *EMBO J.*, **4**, 2017.
50. Busslinger, M., Moschonas, N., and Flavell, R. A. (1981). *Cell*, **27**, 289.
51. Maniatis, T., Fritsch, E. F., and Sambrook, J. (ed.) (1982). *Molecular Cloning, A Laboratory Manual*. Cold Spring Harbor Laboratory, NY.
52. Linial, M., Ginderson, N., and Groudine, M. (1985). *Science*, **230**, 1126.
53. Melton, D. A., Krieg, P. A., Rebagliati, M. R., Maniatis, T., Zinn, K., and Green, M. R. (1984). *Nucleic Acids Res.*, **12**, 7035.
54. Maxwell, I. H., Maxwell, E., and Mahn, W. E. (1977). *Nucleic Acids Res.*, **4**, 241.
55. Wright, S. and Bishop, J. M. (1989). *Proc. Natl Acad. Sci. U.S.A.*, **86**, 505.
56. Collis, P., Antoniou, M., and Grosveld, F. (1990). *EMBO J.*, **9**, 233.
57. Soviki, R. K., Scharf, S., Faloona, F., Mullis, K. B., Horn, G. T., Erlich, H. A., and Arnheim, N. (1985). *Science*, **230**, 1350.
58. Koopman, P., Münsterberg, A., Capel, B., Vivian, N., and Lovell-Badge, R. (1990). *Nature*, **348**, 450.
59. McPherson, M. J., Quirke, P., and Taylor, G. R. (ed.) (1991). *PCR: A Practical Approach*. Oxford University Press, Oxford, UK.
60. Brown, T. A. (ed.) (1991). *Essential Molecular Biology: A Practical Approach*. Oxford University Press, Oxford, UK.

6

Chromosome analysis and sorting by flow cytometry

S. MONARD, L. KEARNEY, and B. D. YOUNG

1. Introduction

Flow cytometry can be used to analyse and sort large numbers of chromosomes in a short period with a high reproducibility (review 1). Since its introduction (2, 3) this technique has been refined and developed and now provides an essential resource for the analysis of the human genome. In contrast to conventional slide-based microscopy, chromosomes are isolated as a suspension, stained with one or more fluorescent DNA dyes, and passed singly through one or two laser beams. The intensity of the fluorescent signal from each chromosome is recorded, the values being dependent on the DNA content of each chromosome. Bivariate analysis exploits the base pair binding preferences of DNA specific dyes, usually Hoechst 33258 which has an adenine/thymidine (AT) binding preference in combination with chromomycin A3 which has a guanine/cytosine (GC) binding preference. The intensity of staining with these fluorochromes is dependent not only on the DNA content but is influenced by the base composition of each chromosome.

The data from a single laser machine are often presented in the form of a histogram of number of events versus fluorescence values. With a dual laser system, however, the data accumulated can be presented as a contour map, dot plot, or isometric display. Such flow analysis allows the frequency of different chromosomes to be accurately measured. For example, trisomy 21 would be seen as a 50% increase in number of events in the chromosome 21 peak. The presence of an abnormal chromosome may also be determined provided there is a difference in either DNA content or AT:GC ratio between the abnormal chromosome and the normal counterpart. The accuracy of the flow karyotype is such that there can be differences in both DNA content and AT:GC ratio between the two homologues of the same chromosome in a normal individual. Some abnormalities such as the t(9;22) translocation found in chronic mylogenous leukaemia and trisomy 21 found in Down's syndrome can be detected easily by flow cytometry. Examples of bivariate flow karyotypes are shown in *Figure 1* and the difference between a normal karyotype

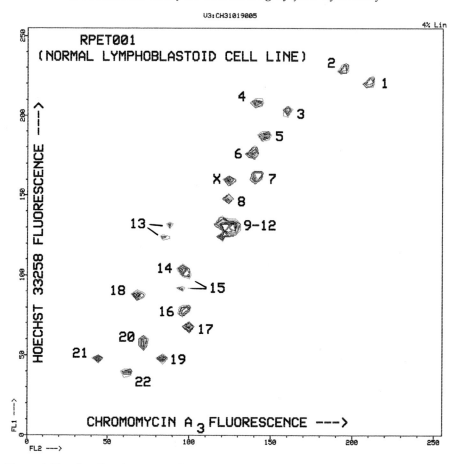

U3:CH31019005

Figure 1 Bivariate flow karyotypes from lymphoid cell lines. (a) Lymphoblastoid cell line RPET001. (b) The Daudi cell line derived from Burkitt's lymphoma. The products of the t(8;14) translocation are indicated.

and an abnormal karyotype can be easily seen. In addition to flow karyotyping, flow cytometry offers the possibility of physically separating individual chromosomes. Phage libraries (4) and more recently cosmid libraries (5) have been constructed from flow sorted chromosomes. The construction of gene libraries requires relatively large numbers of chromosomes, $2-5 \times 10^6$ typically which can take one week or longer to sort.

The development of the polymerase chain reaction (PCR) (6) has opened up the possibility of performing molecular studies on much smaller numbers of chromosomes (7). For example, a method has been described recently (8) for the generation of a library of chromosome specific probes from small numbers (about 500) of flow sorted chromosomes. Small numbers of flow sorted chromosomes can also be used to generate chromosome specific

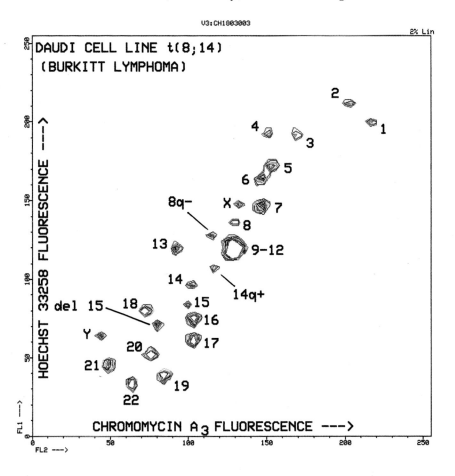

'paints' after Alu-mediated PCR. In this approach the amplified DNA fragments are biotinylated and used to perform *in situ* hybridization on metaphase spreads. The resultant fluorescence signal has been shown to be restricted to the original sorted chromosome (9). Examples are shown in *Figure 2* for paints of chromosomes 2 and 8 generated by flow sorting followed by Alu-primed PCR amplification. This technique potentially can detect chromosome changes which are undetectable by either conventional or flow cytogenetics. The presence or absence of a particular DNA sequence in different chromosomes can also be assessed by PCR with primers specific to that sequence. Our experience suggests that sorting directly into PCR buffer gives the most consistent amplifications

a

b

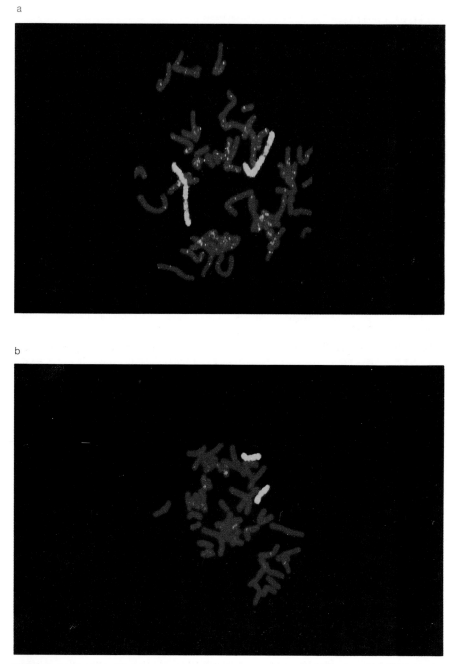

Figure 2 *In situ* hybridization of paints for chromosomes #2 (a) and #8 (b). Paints were generated by flow sorting 200 chromosomes into PCR buffer followed by Alu-PCR amplification.

2. Instrumentation

Univariate chromosome analysis can be performed on most flow cytometers equipped with an argon ion laser (10). Chromosomes are usually stained with ethidium bromide which can be excited with the 514 nm line, or more usually with the 488 nm line of an argon ion laser. The emission signal can be collected using a 580 nm long pass filter or equivalent band pass filter. Other fluorochromes can be used for univariate analysis and sorting if they give a clearer separation of the chromosome peak of interest.

A flow cytometer equipped with two 5 watt argon ion lasers is required for bivariate analysis of chromosomes stained with Hoechst 33258 and chromomycin A3 (11). The primary laser, i.e. the laser that intersects the sample stream nearest the nozzle, is tuned to the UV lines from 351.1 to 363.8 nm and is used to excite the Hoechst dye. The secondary laser is tuned to 457.9 nm and is used to excite the chromomycin. A 390 nm long pass and a 480 nm short pass filter are used to collect the signal from the Hoechst dye, and the chromomycin signal is collected using a 490 nm long pass filter only. All are of the coloured glass variety. The signal from the primary laser is usually designated fluorescence 1 (FL1) and the signal from the secondary laser is usually designated fluorescence 2 (FL2). Examples of bivariate flow karyotypes are shown in *Figure 1*.

The alignment of the laser beams is of critical importance. Special care should be taken to ensure that they do not pass too close to the edge of any of the prisms as this can cause diffraction and subsequent loss of resolution. It is important that the lasers are functioning in the TEM00 mode. The condition of the laser focusing lens should be inspected periodically. Acromatic doublets, which are two lenses of different matcrials positioned close to each other or cemented together, are usually used. If the lens is of the cemented type, the cement may discolour after prolonged use in the UV. The collection filters, like all the optical components, should be kept scrupulously clean. Fluorescent microspheres are useful for aligning the sample stream and collection optics. With a dual laser system, the secondary laser should be blocked while the sample stream position, collection optics, and laser focusing are adjusted to give the sharpest peak from the microspheres in the FL1 channel. The secondary laser can then be unblocked and it should be possible to get a signal with a low CV from the microspheres on the FL2 channel by adjusting the secondary laser beam with minimum change in position of the sample stream, collection optics, or laser focusing lens.

The fluidic system must be kept clean. Sheath filters should be changed regularly and fluid resistors should be flushed through with a detergent solution periodically. The usual nozzle orifice size used is 50 μm although a 70 μm nozzle can be used. A dirty nozzle can cause poor resolution, increased noise, and irregular deflection streams. The nozzle orifice can be inspected by placing orifice down on a microscope slide and viewing on an inverted microscope.

Sheath buffer should be autoclaved before use to destroy any DNase activity present. When sorting chromosomes, particularly for library construction, keeping the sample and sorted fraction cool with circulating cold water reduces the chance of DNA degradation.

The only parameters necessary for bivariate chromosome analysis are the two fluorescence parameters but it can be useful to use one or more scatter parameters for live gating. For this purpose the obscuration bar of the forward scatter detector may have to be slightly enlarged. The instrument should be triggered on the Hoechst signal and it is not advisable to use anything but the minimum possible threshold, as a high threshold would allow small particles such as chromosome fragments to pass undetected through the instrument and into the sorted droplets.

It is advisable to do a test sort before commencing any chromosome sorting. This can be done on fluorescent microspheres or preferably on chromosomes. The sorted chromosomes can either be restained with Hoechst and chromomycin A3 and rerun on the same instrument or if a bench-top analyser such as a FACSCAN, is available, the chromosomes can be stained with propidium iodide and run on this instrument. The sample must be allowed to equilibrate with the instrument tubing by running it on the cytometer for 10–30 minutes. After this period the profile should have stabilized and sort windows can be set.

Sorting chromosomes for library construction is a time consuming process. With a good preparation and a well set-up instrument it should be possible to pass 1500–2500 chromosomes through the instrument per second without deterioration of resolution. Thus it should be possible to sort a single copy chromosome at a rate of 30–50 per second. When running chromosomes for analysis the best discrimination between chromosome peaks will be obtained using a low sample rate.

2.1 Safety

Care should be exercised when aligning the lasers especially in the UV. The lowest laser powers should be used and protective goggles should be worn at all times.

Many of the reagents used in the preparation and staining of chromosomes are highly toxic or potentially mutagenic, the use of disposable gloves is advisable. This may also prevent possible contamination from nucleases.

Protocol 1. Polyamine method (basic protocol)

Materials and equipment
- tissue culture flasks
- 50 ml conical tubes

- complete media, suitable for chosen cells
- bench centrifuge
- fluorescence microscope
- hotplate/magnetic stirrer
- colcemid (Gibco/BRL)
- incubator
- hypotonic swelling solution (see *Table 1*)
- chromosome isolation buffer 1 (CIB1), 10 × stock solution (see *Table 1*)
- digitonin (Sigma)
- 12 × 75 mm plastic tubes
- propidium iodide 50 µg/ml in phosphate-buffered saline (PBS)
- propidium iodide 50 µg/ml in PBS + 0.1% Triton X-100

1. Cell lines, either monolayer or suspension may be used, as may phyto-haemagglutinin stimulated peripheral blood cells. Whichever type of cell is used, the best preparations will be made from healthy cells growing optimally. Cell lines must be mycoplasma free and preferably split 24 h before blocking with colcemid.

2. Cells should be blocked with 0.05 µg/ml colcemid for 5–16 h depending on rate of cell growth. Usually blocking overnight gives good results. The proportion of cells in mitosis can be estimated by pelleting the cells from 1 ml of the blocked cell culture, discarding the supernatant, and re-suspending in phosphate-buffered saline (PBS) containing 50 µg/ml propidium iodide (PI) and 0.1% Triton X-100. The estimation can be made either using a fluorescence microscope or a bench-top flow cytometer. It should be possible to get 40–60% of the cells in mitosis with suspension cell lines.

3. Mitotic shake-off can be used with monolayer cell lines. The mitotic cells are selectively shaken-off and released into the media, and can be spun down in 50 ml tubes at 100 *g* for 10 min. With suspension cell lines, the cells are spun at 100 *g* for 10 min. Discard the supernatant, resuspend the cells in fresh medium, and spin again for 10 min at 100 *g*.

4. Discard the supernatant by inverting the tube. The last few drops can be removed with a tissue. Flick the tube gently until the pellet becomes loose. Add to about ten times the volume of the pellet (hypotonic swelling solution), mix gently, and leave for 10–30 min at room temperature. Lymphoblastoid cell lines usually require about 20 min in swelling buffer. The contents of several tubes can be pooled. Swelling can be monitored microscopically. Spin at 100 *g* for 10 min.

Protocol 1. *Continued*

5. Meanwhile, prepare the chromosome isolation buffer (CIB1) by adding 1 ml of 10 × CIB1, 9 ml of distilled water, and 12 mg of digitonin together with a small magnetic 'flea' to a glass Universal, warm gently on the hotplate stirrer until the digitonin has dissolved. Allow to cool to room temperature then adjust the pH to 7.2 using 0.1 M sodium hydroxide. Filter through a 0.22 μm sterile disposable filter into a new tube and place on ice. It is preferable to make the chromosome suspensions very concentrated, it is not necessary to use more than 10 ml of CIB for 200 ml of suspension cell culture.

6. Carefully remove the supernatant with a Pasteur pipette and flick the tube gently. The ice-cold CIB should now be added to about five times the volume of the pellet, flick the tube gently. Mix a small amount of the preparation with an equal volume of PBS with 50 μg/ml PI on a microscope slide, put on a coverslip, and view with a fluorescence microscope. If the chromosomes are not monodispersed the preparation can be flicked more vigorously or sucked up and down a plastic Pasteur pipette. Vortexing can result in chromosome breakage and should be avoided if possible.

7. The chromosome suspension may be stored at 4°C for several weeks with little deterioration of flow karyotype.

8. Transfer 1.2 ml of the chromosome suspension into a 12 × 75 mm plastic test-tube, centrifuge at 200 *g* for 1 min to remove most of the intact nuclei. Transfer 800 μl of the supernatant into a new tube. Add 30 μl Hoechst 33258 (100 μg/ml in dH_2O), mix immediately. Add 40 μl of 15 mM $MgSO_4$ and 50 μl chromomycin A3 (2 mg/ml in ethanol). Mix and leave the sample at 4°C for 2 h in the dark.

9. The chromosome profile can be improved if 100 μl sodium citrate (100 mM) and 100 μl sodium sulphite (250 mM) are added 15 min prior to running on cytometer.

Protocol 2. Magnesium sulphate method (alternative protocol)

Materials and equipment
- chromsome isolation buffer 2 (CIB2) (see *Table 1*)
- Triton X-100
- tissue culture flasks
- 50 ml conical tubes
- 12 × 75 mm plastic tubes
- complete media, suitable for chosen cells

- bench centrifuge
- fluorescence microscope
- incubator
- vortex mixer
- colcemid (Gibco/BRL)
- phase contrast microscope

1. Prepare colcemid blocked cells as in *Protocol 1*.
2. Centrifuge cells at 300 *g* for 10 min at room temperature, decant supernatant, draining tubes on an absorbent paper towel.
3. Add 1 ml of CIB2 to 6×10^5 cells, resuspend gently, and incubate at room temperature for 10 min.
4. Add 0.1 ml of Triton X-100 solution (2.5% in distilled water) and incubate on ice for 10 min.
5. Vortex for 10–20 sec to disrupt the cells (monitor using phase contrast microscopy).
6. Stain for bivariate analysis as in step **8** of *Protocol 1*.

Protocol 3. Preparation of chromosomes for library construction (supplementary protocol)

Materials and equipment
- sheath buffer (see *Table 1*)
- tRNA (BRL)
- proteinase K (BDH)
- sterile 1.5 ml conical tubes with screw caps
- stock solution 500 mM EDTA pH 8.0
- stock solution 20% w/v *n*-lauroylsarcosine (sodium salt)

1. Sterile sheath buffer should be prepared containing 500 μg/ml tRNA. 50 μl of this solution should be dispensed into sterile 1.5 ml conical tubes and vortexed vigorously to coat the inside of the tubes.
2. The tubes can be stored by freezing quickly on dry-ice and stored at −20 °C.
3. A working solution of 250 mM EDTA and 10% w/v *n*-lauroylsarcosine should be made. This is added to the sorted chromosome suspension to make a final concentration of 25 mM EDTA and 1% *n*-lauroylsarcosine. When using a 50 μm nozzle and a three drop deflection 5×10^5 chromosomes should occupy a volume of about 700–800 μl, and thus 70–80 μl working solution should be added.

Protocol 3. *Continued*

4. 180 μg of proteinase K is added, the tubes are vortexed, and incubated at 42°C overnight.

5. The resulting DNA preparation can be stored at 4°C for many months.

Table 1. Reagents, solutions, and media

Chromosome isolation buffer1

NaCl	20 mM
KCl	80 mM
Tris–HCl	15 mM
EGTA	0.5 mM
EDTA	2 mM
β-mercaptoethanol	0.15% w/v
spermine	0.2 mM
spermidine	0.5 mM
autoclaved distilled water	

It is convenient to make a 10 × concentrate of this buffer (pH 7.2) and store at 4°C. Before use dilute, dissolve 0.12% w/v digitonin, and adjust to pH 7.2 before filtering through a sterile disposable filter pore size 0.22 μm. Aliquoting 12 mg of digitonin into a number of glass Universals saves time and reduces exposure to this toxic substance.

Chromosome isolation buffer 2

KCl	40 mM
Hepes	5 mM
MgSO$_4$	10 mM
dithiothreitol	3 mM
autoclaved distilled water	

Adjust to pH 8.0 with 0.5 M KOH. Before use, filter through sterile disposable filter of pore size 0.22 μm.

Hypotonic swelling solution

KCl	50 mM
spermine	0.2 mM
spermidine	0.5 mM
distilled water	

Autoclave and filter through a 0.22 μm sterile disposable filter before use.

Sheath buffer

NaCl	100 mM
Tris–HCl	10 mM
EDTA	1 mM
distilled water	

Adjust to pH 8.0 with 10 M NaOH. Suggest decanting into 2 litre glass bottles and autoclaving before use.

---

3. Discussion

Although several different preparative techniques have been developed, they all have certain common features. A culture of growing cells is usually treated with an agent such as colcemid or vinblastine in order to arrest sufficient cells in metaphase. The period of treatment may depend on the cell cycle time, but extensive culture in the presence of such agents can lead first to a high degree of chromosome contraction and ultimately to cell death. If the cells grow as an attached layer, mitotic shake-off can be used to obtain an enriched population of metaphase cells. Suspension cell lines may also be used if sufficient metaphase cells are present and if care is taken not to lyse nuclei of the interphase cells which predominate in the population. The cell population is usually subjected to hypotonic swelling, treatment with a detergent, and lysis by mechanical disruption such as passage through a fine needle or by vigorous vortexing. During the lysis process it is helpful to monitor the disruption of the mitotic cells on a fluorescence or phase contrast microscopy, since insufficient treatment will not lyse sufficient cells or disrupt chromosome clumps whereas excessive treatment can cause chromosome breakage. Differential centrifugation may be used to remove interphase nuclei, if present.

The choice of method for preparing chromosomes for flow cytometry depends on the aims of the particular experiment. The hexalene glycol method (12) is reported to allow banding and hence identification of the chromosomes following sorting but the discrimination of the different chromosome types on a commercially available flow cytometer is relatively poor. The preparative technique in routine use in our laboratory is described in *Protocol 1*. This method, which uses polyamines to stabilize the chromosomal DNA (13) has the particular advantage that very high molecular weight DNA can be obtained after sorting (14). The discrimination of the different chromosomes is good making it an ideal method to use when sorting chromosomes for cosmid library construction. The method described in *Protocol 2* (11) is thought to give the best discrimination between chromosome types, although it may not give such high molecular weight DNA as the polyamine method. Any damage to chromosomes not only alters their DNA content but may cause aggregation and loss of resolution thereby resulting in chromosome fragments or changes that cannot be distinguished from normal chromosomes. These 'debris particles' produce a slowly varying continuum in the flow karyotype that underlies the peaks produced by intact chromosomes. Hence the ideal technique would optimize cell lysis and minimize chromosome damage.

The choice of stain for chromosome sorting and analysis depends largely on the configuration of the flow cytometer and in particular on the light source available. Bivariate analysis using Hoechst 33258 and chromomycin A3 offers the clearest separation of the different chromosome types (*Figure 1*). If a single laser cytometer is to be used there are three main choices of DNA stain. Commonly used is ethidium bromide or propidium iodide, and these dyes can be excited using a 488 nm light source and are thought to have no base preference. Secondly, one of the UV excited dyes such as Hoechst 33258 or DAPI can be used. These dyes have an AT binding preference. Lastly, chromomycin or mithromycin can be used, these dyes have a GC binding preference and can be excited by the 457.9 nm line of an argon laser. Ethidium bromide is not always the most suitable dye for chromosome analysis and sorting, for example, chromosomes 1 and 2 can often be more readily discriminated, and hence sorted, using chromomycin A3.

It is not always easy to determine if the poor discrimination between chromosomes seen on the cytometer is due to a poor preparation or poor cytometer performance. The performance of the instrument can be monitored using fluorescent microspheres. The CVs of the fluorescent peaks and the intensity of the signal give an indication of how the instrument is performing. Staining the preparation with propidium iodide and viewing with a fluorescent microscope will show whether there are too few chromosomes or the chromosomes are aggregating. Finally, it is invaluable to compare the results obtained with those of another laboratory where chromosome sorting or analysis is constantly performed.

The practical limit to the resolution obtained in flow karyotypes is an aggregate of many factors. First, the condition of the chromosome preparation is most important since damage or aggregation can cause a severe loss of resolution. The fluorochrome must be used under saturation conditions such that the fluorescence intensity is independent of chromosome condensation. Finally, the optical and flow settings of the cytometer must be optimal for maximum resolution. For example, a substantial increase in resolution can often be seen when the sample flow rate is reduced causing the chromosomes to be more centrally aligned in the liquid stream. The use of fluorescent microspheres can also be very helpful in finding the optimal machine settings prior to analysis of the chromosome sample.

Mathematical analysis of flow karyotypes is best achieved by least squares fitting of Gaussian distributions to the raw data (15, 16). Such analysis has shown that single parameter flow karyotypes can be achieved with coefficients of variation of 1–2% (10). At this level of resolution it should be possible to detect differences in chromosomal DNA content of 1/2000th of the male genome. This represents less than one prometaphase band and therefore flow cytogenetics is potentially useful for the detection of chromosomal aberrations including small alterations in DNA content. For example, a trisomy is apparent as a 50% increase in area of the relevant peak. A balanced reciprocal translocation between two chromosomes, with a net change in their DNA contents, results in two new peaks whose area is each equivalent to a single chromosome and the peaks due to the two unaffected homologues should be reduced to that of single chromosome. An aberration such as a small marker chromosome should result in a peak in a position where normally no peaks exist.

In practice the detection of flow karyotype abnormalities without the support of conventional cytogenetics would be greatly complicated by the high level of polymorphism in the general population. Certain chromosomes have regions of centric heterochromatin which can vary considerably in size, resulting in microscopically visible differences in size (17). Variations in flow karyotypes have been correlated with specific C or Q band heteromorphisms (10, 18). These variations were investigated further (19) where chromosomal DNA content was determined from a series of 20 normal individuals using high resolution flow cytometry. The mean relative fluorescence values for each chromosome are in close agreement with cytophotometric values of DNA content (20). It was clear that chromosomes 1, 9, 16, and Y showed the largest variations, in accord with their large centric heterochromatic regions. Chromosome heteromorphisms appear to be inherited unaltered in size (21) and hence any feature which cannot be seen in either of the parental flow karyotypes can be assumed to be of *de novo* occurrence (22). This approach has been used (23) to study a series of families with dysmorphic children. Although no chromosome abnormality was detected the usefulness of this approach was clearly demonstrated.

Bivariate flow karyotyping has been used in the classification of chromosomes with a view to the detection of numerical and structural aberrations (24–26). In a particular study (26) of amniocentesis samples cytogenetic information determined on the basis of flow karyotypes was compared with that obtained by visual analysis following G-banding. Numerical aberrations involving chromosomes 21, 18, and Y were detected correctly in all of 28 analyses, including eight in a blind study.

References

1. Young, B. D. (1990). In *Flow Cytometry: A Practical Approach* (ed.) M. G. Ormerod, pp. 145–59. IRL Press, Oxford.
2. Gray, J. W., Carrano, A. V., Steinmetz, L. L., Van Dilla, M. A., Moore II, D. H., Mayall, B. H., and Mendelsohn, M. L. (1975). *Proc. Natl Acad. Sci. U.S.A.*, **72**, 1231.
3. Gray, J. W., Carrano, A. V., Moore II, D. H., Steinmetz, L. L., Minkler, J., Mayall, B. H., Mendelsohn, M. L., and Van Dilla, M. A. (1975). *Clin. Chem.*, **21**, 1258.
4. Krumlauf, R., Jeanpierre, M., and Young, B. D. (1982). *Proc. Natl Acad. Sci. U.S.A.*, **79**, 2971.
5. Nizetic, D., Zehetner, G., Monaco, A. P., Young, B. D., and Lehrach, H. (1991). *Proc. Natl Acad. Sci. U.S.A.*, **88**, 3233.
6. Saiki, R. K., Gelfand, D. J., Stoffel, S., Scharf, S. J., Higuchi, R., Horn, G. T., Mullis, K. B., and Erlich, H. A. (1988). *Science*, **239**, 487.
7. Cotter, F., Nasipuri, S., Lam, G., and Young, B. D. (1989). *Genomics*, **5**, 470.
8. Cotter, F. E., Das, S., Douek, E., Carter, N. P., and Young, B. D. (1991). *Genomics*, **9**, 473.

9. Suijkerbuijk, R. F., Matthopoulos, D., Kearney, L., Monard, S., Dhut, S., Cotter, F. E., Herbergs, J., Geurts van Kessel, A., and Young, B. D. (1992). *Genomics*, **13**, 355.
10. Young, B. D., Ferguson-Smith, M. A., Sillar, R., and Boyd, E. (1981). *Proc. Natl Acad. Sci. U.S.A.*, **78**, 7727.
11. van den Engh, G. J., Trask, B., Gray, J. W., Langlois, R. G., and Yu, L. C. (1985). *Cytometry*, **6**, 92.
12. Stubblefield, E., Cram, S., and Deaven, L. (1975). *Exp. Cell Res.*, **94**, 464.
13. Sillar, R. and Young, B. D. (1981). *J. Histochem. Cytochem.*, **29**, 74.
14. Minoshima, S., Kawasaki, K., Fukuyama, R., Maekawa, M., Kudoh, J., and Shimizu, N. (1990). *Cytometry*, **11**, 539.
15. Dean, P. N., Kolla, S., and Van, D. M. (1989). *Cytometry*, **10**, 109.
16. van den Engh, G., Hanson, D., and Trask, B. (1990). *Cytometry*, **11**, 173.
17. Trask, B., van den Engh, G., Mayall, B., and Gray, J. W. (1989). *Am. J. Hum. Genet.*, **45**, 739.
18. Langlois, R. G., Yu, L. C., Gray, J. W., and Carrano, A. V. (1982). *Exp. Cell Res.*, **133**, 341.
19. Harris, P., Boyd, E., Young, B. D., and Ferguson-Smith, M. A. (1986). *Cytogenet.. Cell Genet.*, **41**, 14.
20. Mayall, B. H., Carrano, A. V., Moore, D. H., Ashworth, L., Bennet, D., and Mendelsohn, M. (1984). *Cytometry*, **9**, 376.
21. Robinson, J. A., Buckton, K. E., Spowart, G., Newton, M., Jacobs, P. A., Evans, H. J., and Hill, R. (1986). *Am. J. Hum. Genet.*, **40**, 113.
22. Trask, B., van den Engh, G., and Gray, J. W. (1989). *Am. J. Hum. Genet.*, **45**, 753.
23. Harris, P., Cooke, A., Boyd, E., Young, B. D., and Ferguson-Smith, M. A. (1987). *Hum. Genet.*, **76**, 129.
24. Carter, N. P., Ferguson-Smith, M. E., Affara, N. A., Briggs, H., and Ferguson-Smith, M. A. (1990). *Cytometry*, **11**, 202.
25. Trask, B., van den Engh, G., Nussbaum, R., Schwartz, C., and Gray, J. (1990). *Cytometry*, **11**, 184.
26. Gray, J. W., Trask, B., van den Engh, G., Silva, A., Lozes, C., Grell, S., Schonberg, S., Yu, L. C., and Golbus, M. S. (1988). *Am. J. Hum. Genet.*, **42**, 49.

Index

ORDER OTHER TITLES OF INTEREST TODAY

Price list for: UK, Europe, Rest of World (excluding US and Canada)

Forthcoming Titles

124. Human Genetic Disease Analysis Davies, K.E. (Ed)
...... Spiralbound hardback 0-19-963309-6 **£30.00**
...... Paperback 0-19-963308-8 **£18.50**
123. Protein Phosphorylation Hardie, G. (Ed)
...... Spiralbound hardback 0-19-963306-1 **£32.50**
...... Paperback 0-19-963305-3 **£22.50**
122. Immunocytochemistry Beesley, J. (Ed)
...... Spiralbound hardback 0-19-963270-7 **£32.50**
...... Paperback 0-19-963269-3 **£22.50**
121. Tumour Immunobiology Gallagher, G., Rees, R.C. & others (Eds)
...... Spiralbound hardback 0-19-963370-3 **£35.00**
...... Paperback 0-19-963369-X **£25.00**
120. Transcription Factors Latchman, D.S. (Ed)
...... Spiralbound hardback 0-19-963342-8 **£30.00**
...... Paperback 0-19-963341-X **£19.50**
119. Growth Factors McKay, I.A. & Leigh, I. (Eds)
...... Spiralbound hardback 0-19-963360-6 **£30.00**
...... Paperback 0-19-963359-2 **£19.50**
118. Histocompatibility Testing Dyer, P. & Middleton, D. (Eds)
...... Spiralbound hardback 0-19-963364-9 **£32.50**
...... Paperback 0-19-963363-0 **£22.50**
117. Gene Transcription Hames, D.B. & Higgins, S.J. (Eds)
...... Spiralbound hardback 0-19-963292-8 **£35.00**
...... Paperback 0-19-963291-X **£25.00**
116. Electrophysiology Wallis, D.I. (Ed)
...... Spiralbound hardback 0-19-963348-7 **£32.50**
...... Paperback 0-19-963347-9 **£22.50**
115. Biological Data Analysis Fry, J.C. (Ed)
...... Spiralbound hardback 0-19-963340-1 **£50.00**
...... Paperback 0-19-963339-8 **£27.50**
114. Experimental Neuroanatomy Bolam, J.P. (Ed)
...... Spiralbound hardback 0-19-963326-6 **£32.50**
...... Paperback 0-19-963325-8 **£22.50**
112. Lipid Analysis Hamilton, R.J. & Hamilton, S.J. (Eds)
...... Spiralbound hardback 0-19-963098-4 **£35.00**
...... Paperback 0-19-963099-2 **£25.00**
111. Haemopoiesis Testa, N.G. & Molineux, G. (Eds)
...... Spiralbound hardback 0-19-963366-5 **£32.50**
...... Paperback 0-19-963365-7 **£22.50**

Published Titles

113. Preparative Centrifugation Rickwood, D. (Ed)
...... Spiralbound hardback 0-19-963208-1 **£45.00**
...... Paperback 0-19-963211-1 **£25.00**
110. Pollination Ecology Dafni, A.
...... Spiralbound hardback 0-19-963299-5 **£32.50**
...... Paperback 0-19-963298-7 **£22.50**
109. In Situ Hybridization Wilkinson, D.G. (Ed)
...... Spiralbound hardback 0-19-963328-2 **£30.00**
...... Paperback 0-19-963327-4 **£18.50**
108. Protein Engineering Rees, A.R., Sternberg, M.J.E. & others (Eds)
...... Spiralbound hardback 0-19-963139-5 **£35.00**
...... Paperback 0-19-963138-7 **£25.00**

107. Cell-Cell Interactions Stevenson, B.R., Gallin, W.J. & others (Eds)
...... Spiralbound hardback 0-19-963319-3 **£32.50**
...... Paperback 0-19-963318-5 **£22.50**
106. Diagnostic Molecular Pathology: Volume I Herrington, C.S. & McGee, J. O'D. (Eds)
...... Spiralbound hardback 0-19-963237-5 **£30.00**
...... Paperback 0-19-963236-7 **£19.50**
105. Biomechanics-Materials Vincent, J.F.V. (Ed)
...... Spiralbound hardback 0-19-963223-5 **£35.00**
...... Paperback 0-19-963222-7 **£25.00**
104. Animal Cell Culture (2/e) Freshney, R.I. (Ed)
...... Spiralbound hardback 0-19-963212-X **£30.00**
...... Paperback 0-19-963213-8 **£19.50**
103. Molecular Plant Pathology: Volume II Gurr, S.J., McPherson, M.J. & others (Eds)
...... Spiralbound hardback 0-19-963352-5 **£32.50**
...... Paperback 0-19-963351-7 **£22.50**
101. Protein Targeting Magee, A.I. & Wileman, T. (Eds)
...... Spiralbound hardback 0-19-963206-5 **£32.50**
...... Paperback 0-19-963210-3 **£22.50**
100. Diagnostic Molecular Pathology: Volume II: Cell and Tissue Genotyping Herrington, C.S. & McGee, J.O'D. (Eds)
...... Spiralbound hardback 0-19-963239-1 **£30.00**
...... Paperback 0-19-963238-3 **£19.50**
99. Neuronal Cell Lines Wood, J.N. (Ed)
...... Spiralbound hardback 0-19-963346-0 **£32.50**
...... Paperback 0-19-963345-2 **£22.50**
98. Neural Transplantation Dunnett, S.B. & Björklund, A. (Eds)
...... Spiralbound hardback 0-19-963286-3 **£30.00**
...... Paperback 0-19-963285-5 **£19.50**
97. Human Cytogenetics: Volume II: Malignancy and Acquired Abnormalities (2/e) Rooney, D.E. & Czepulkowski, B.H. (Eds)
...... Spiralbound hardback 0-19-963290-1 **£30.00**
...... Paperback 0-19-963289-8 **£22.50**
96. Human Cytogenetics: Volume I: Constitutional Analysis (2/e) Rooney, D.E. & Czepulkowski, B.H. (Eds)
...... Spiralbound hardback 0-19-963288-X **£30.00**
...... Paperback 0-19-963287-1 **£22.50**
95. Lipid Modification of Proteins Hooper, N.M. & Turner, A.J. (Eds)
...... Spiralbound hardback 0-19-963274-X **£32.50**
...... Paperback 0-19-963273-1 **£22.50**
94. Biomechanics-Structures and Systems Biewener, A.A. (Ed)
...... Spiralbound hardback 0-19-963268-5 **£42.50**
...... Paperback 0-19-963267-7 **£25.00**
93. Lipoprotein Analysis Converse, C.A. & Skinner, E.R. (Eds)
...... Spiralbound hardback 0-19-963192-1 **£30.00**
...... Paperback 0-19-963231-6 **£19.50**
92. Receptor-Ligand Interactions Hulme, E.C. (Ed)
...... Spiralbound hardback 0-19-963090-9 **£35.00**
...... Paperback 0-19-963091-7 **£25.00**
91. Molecular Genetic Analysis of Populations Hoelzel, A.R. (Ed)
...... Spiralbound hardback 0-19-963278-2 **£32.50**
...... Paperback 0-19-963277-4 **£22.50**

90. **Enzyme Assays** Eisenthal, R. & Danson, M.J. (Eds)
...... Spiralbound hardback 0-19-963142-5 **£35.00**
...... Paperback 0-19-963143-3 **£25.00**
89. **Microcomputers in Biochemistry** Bryce, C.F.A. (Ed)
...... Spiralbound hardback 0-19-963253-7 **£30.00**
...... Paperback 0-19-963252-9 **£19.50**
88. **The Cytoskeleton** Carraway, K.L. & Carraway, C.A.C. (Eds)
...... Spiralbound hardback 0-19-963257-X **£30.00**
...... Paperback 0-19-963256-1 **£19.50**
87. **Monitoring Neuronal Activity** Stamford, J.A. (Ed)
...... Spiralbound hardback 0-19-963244-8 **£30.00**
...... Paperback 0-19-963243-X **£19.50**
86. **Crystallization of Nucleic Acids and Proteins** Ducruix, A. & Giegꞇ130ꞈ, R. (Eds)
...... Spiralbound hardback 0-19-963245-6 **£35.00**
...... Paperback 0-19-963246-4 **£25.00**
85. **Molecular Plant Pathology: Volume I** Gurr, S.J., McPherson, M.J. & others (Eds)
...... Spiralbound hardback 0-19-963103-4 **£30.00**
...... Paperback 0-19-963102-6 **£19.50**
84. **Anaerobic Microbiology** Levett, P.N. (Ed)
...... Spiralbound hardback 0-19-963204-9 **£32.50**
...... Paperback 0-19-963262-6 **£22.50**
83. **Oligonucleotides and Analogues** Eckstein, F. (Ed)
...... Spiralbound hardback 0-19-963280-4 **£32.50**
...... Paperback 0-19-963279-0 **£22.50**
82. **Electron Microscopy in Biology** Harris, R. (Ed)
...... Spiralbound hardback 0-19-963219-7 **£32.50**
...... Paperback 0-19-963215-4 **£22.50**
81. **Essential Molecular Biology: Volume II** Brown, T.A. (Ed)
...... Spiralbound hardback 0-19-963112-3 **£32.50**
...... Paperback 0-19-963113-1 **£22.50**
80. **Cellular Calcium** McCormack, J.G. & Cobbold, P.H. (Eds)
...... Spiralbound hardback 0-19-963131-X **£35.00**
...... Paperback 0-19-963130-1 **£25.00**
79. **Protein Architecture** Lesk, A.M.
...... Spiralbound hardback 0-19-963054-2 **£32.50**
...... Paperback 0-19-963055-0 **£22.50**
78. **Cellular Neurobiology** Chad, J. & Wheal, H. (Eds)
...... Spiralbound hardback 0-19-963106-9 **£32.50**
...... Paperback 0-19-963107-7 **£22.50**
77. **PCR** McPherson, M.J., Quirke, P. & others (Eds)
...... Spiralbound hardback 0-19-963226-X **£30.00**
...... Paperback 0-19-963196-4 **£19.50**
76. **Mammalian Cell Biotechnology** Butler, M. (Ed)
...... Spiralbound hardback 0-19-963207-3 **£30.00**
...... Paperback 0-19-963209-X **£19.50**
75. **Cytokines** Balkwill, F.R. (Ed)
...... Spiralbound hardback 0-19-963218-9 **£35.00**
...... Paperback 0-19-963214-6 **£25.00**
74. **Molecular Neurobiology** Chad, J. & Wheal, H. (Eds)
...... Spiralbound hardback 0-19-963108-5 **£30.00**
...... Paperback 0-19-963109-3 **£19.50**
73. **Directed Mutagenesis** McPherson, M.J. (Ed)
...... Spiralbound hardback 0-19-963141-7 **£30.00**
...... Paperback 0-19-963140-9 **£19.50**
72. **Essential Molecular Biology: Volume I** Brown, T.A. (Ed)
...... Spiralbound hardback 0-19-963110-7 **£32.50**
...... Paperback 0-19-963111-5 **£22.50**
71. **Peptide Hormone Action** Siddle, K. & Hutton, J.C.
...... Spiralbound hardback 0-19-963070-4 **£32.50**
...... Paperback 0-19-963071-2 **£22.50**
70. **Peptide Hormone Secretion** Hutton, J.C. & Siddle, K. (Eds)
...... Spiralbound hardback 0-19-963068-2 **£35.00**
...... Paperback 0-19-963069-0 **£25.00**
69. **Postimplantation Mammalian Embryos** Copp, A.J. & Cockroft, D.L. (Eds)
...... Spiralbound hardback 0-19-963088-7 **£35.00**
...... Paperback 0-19-963089-5 **£25.00**
68. **Receptor-Effector Coupling** Hulme, E.C. (Ed)
...... Spiralbound hardback 0-19-963094-1 **£30.00**
...... Paperback 0-19-963095-X **£19.50**

67. **Gel Electrophoresis of Proteins (2/e)** Hames, B.D. & Rickwood, D. (Eds)
...... Spiralbound hardback 0-19-963074-7 **£35.00**
...... Paperback 0-19-963075-5 **£25.00**
66. **Clinical Immunology** Gooi, H.C. & Chapel, H. (Eds)
...... Spiralbound hardback 0-19-963086-0 **£32.50**
...... Paperback 0-19-963087-9 **£22.50**
65. **Receptor Biochemistry** Hulme, E.C. (Ed)
...... Spiralbound hardback 0-19-963092-5 **£35.00**
...... Paperback 0-19-963093-3 **£25.00**
64. **Gel Electrophoresis of Nucleic Acids (2/e)** Rickwood, D. & Hames, B.D. (Eds)
...... Spiralbound hardback 0-19-963082-8 **£32.50**
...... Paperback 0-19-963083-6 **£22.50**
63. **Animal Virus Pathogenesis** Oldstone, M.B.A. (Ed)
...... Spiralbound hardback 0-19-963100-X **£30.00**
...... Paperback 0-19-963101-8 **£18.50**
62. **Flow Cytometry** Ormerod, M.G. (Ed)
...... Paperback 0-19-963053-4 **£22.50**
61. **Radioisotopes in Biology** Slater, R.J. (Ed)
...... Spiralbound hardback 0-19-963080-1 **£32.50**
...... Paperback 0-19-963081-X **£22.50**
60. **Biosensors** Cass, A.E.G. (Ed)
...... Spiralbound hardback 0-19-963046-1 **£30.00**
...... Paperback 0-19-963047-X **£19.50**
59. **Ribosomes and Protein Synthesis** Spedding, G. (Ed)
...... Spiralbound hardback 0-19-963104-2 **£32.50**
...... Paperback 0-19-963105-0 **£22.50**
58. **Liposomes** New, R.R.C. (Ed)
...... Spiralbound hardback 0-19-963076-3 **£35.00**
...... Paperback 0-19-963077-1 **£22.50**
57. **Fermentation** McNeil, B. & Harvey, L.M. (Eds)
...... Spiralbound hardback 0-19-963044-5 **£30.00**
...... Paperback 0-19-963045-3 **£19.50**
56. **Protein Purification Applications** Harris, E.L.V. & Angal, S. (Eds)
...... Spiralbound hardback 0-19-963022-4 **£30.00**
...... Paperback 0-19-963023-2 **£18.50**
55. **Nucleic Acids Sequencing** Howe, C.J. & Ward, E.S. (Eds)
...... Spiralbound hardback 0-19-963056-9 **£30.00**
...... Paperback 0-19-963057-7 **£19.50**
54. **Protein Purification Methods** Harris, E.L.V. & Angal, S. (Eds)
...... Spiralbound hardback 0-19-963002-X **£30.00**
...... Paperback 0-19-963003-8 **£20.00**
53. **Solid Phase Peptide Synthesis** Atherton, E. & Sheppard, R.C.
...... Spiralbound hardback 0-19-963066-6 **£30.00**
...... Paperback 0-19-963067-4 **£18.50**
52. **Medical Bacteriology** Hawkey, P.M. & Lewis, D.A. (Eds)
...... Spiralbound hardback 0-19-963008-9 **£38.00**
...... Paperback 0-19-963009-7 **£25.00**
51. **Proteolytic Enzymes** Beynon, R.J. & Bond, J.S. (Eds)
...... Spiralbound hardback 0-19-963058-5 **£30.00**
...... Paperback 0-19-963059-3 **£19.50**
50. **Medical Mycology** Evans, E.G.V. & Richardson, M.D. (Eds)
...... Spiralbound hardback 0-19-963010-0 **£37.50**
...... Paperback 0-19-963011-9 **£25.00**
49. **Computers in Microbiology** Bryant, T.N. & Wimpenny, J.W.T. (Eds)
...... Paperback 0-19-963015-1 **£19.50**
48. **Protein Sequencing** Findlay, J.B.C. & Geisow, M.J. (Eds)
...... Spiralbound hardback 0-19-963012-7 **£30.00**
...... Paperback 0-19-963013-5 **£18.50**
47. **Cell Growth and Division** Baserga, R. (Ed)
...... Spiralbound hardback 0-19-963026-7 **£30.00**
...... Paperback 0-19-963027-5 **£18.50**
46. **Protein Function** Creighton, T.E. (Ed)
...... Spiralbound hardback 0-19-963006-2 **£32.50**
...... Paperback 0-19-963007-0 **£22.50**
45. **Protein Structure** Creighton, T.E. (Ed)
...... Spiralbound hardback 0-19-963000-3 **£32.50**
...... Paperback 0-19-963001-1 **£22.50**
44. **Antibodies: Volume II** Catty, D. (Ed)
...... Spiralbound hardback 0-19-963018-6 **£30.00**
...... Paperback 0-19-963019-4 **£19.50**

43.	**HPLC of Macromolecules** Oliver, R.W.A. (Ed)		
......	Spiralbound hardback	0-19-963020-8	**£30.00**
......	Paperback	0-19-963021-6	**£19.50**
42.	**Light Microscopy in Biology** Lacey, A.J. (Ed)		
......	Spiralbound hardback	0-19-963036-4	**£30.00**
......	Paperback	0-19-963037-2	**£19.50**
41.	**Plant Molecular Biology** Shaw, C.H. (Ed)		
......	Paperback	1-85221-056-7	**£22.50**
40.	**Microcomputers in Physiology** Fraser, P.J. (Ed)		
......	Spiralbound hardback	1-85221-129-6	**£30.00**
......	Paperback	1-85221-130-X	**£19.50**
39.	**Genome Analysis** Davies, K.E. (Ed)		
......	Spiralbound hardback	1-85221-109-1	**£30.00**
......	Paperback	1-85221-110-5	**£18.50**
38.	**Antibodies: Volume I** Catty, D. (Ed)		
......	Paperback	0-947946-85-3	**£19.50**
37.	**Yeast** Campbell, I. & Duffus, J.H. (Ed)		
......	Paperback	0-947946-79-9	**£19.50**
36.	**Mammalian Development** Monk, M. (Ed)		
......	Hardback	1-85221-030-3	**£30.50**
......	Paperback	1-85221-029-X	**£22.50**
35.	**Lymphocytes** Klaus, G.G.B. (Ed)		
......	Hardback	1-85221-018-4	**£30.00**
34.	**Lymphokines and Interferons** Clemens, M.J., Morris, A.G. & others (Eds)		
......	Paperback	1-85221-035-4	**£22.50**
33.	**Mitochondria** Darley-Usmar, V.M., Rickwood, D. & others (Eds)		
......	Hardback	1-85221-034-6	**£32.50**
......	Paperback	1-85221-033-8	**£22.50**
32.	**Prostaglandins and Related Substances** Benedetto, C., McDonald-Gibson, R.G. & others (Eds)		
......	Hardback	1-85221-032-X	**£32.50**
......	Paperback	1-85221-031-1	**£22.50**
31.	**DNA Cloning: Volume III** Glover, D.M. (Ed)		
......	Hardback	1-85221-049-4	**£30.00**
......	Paperback	1-85221-048-6	**£19.50**
30.	**Steroid Hormones** Green, B. & Leake, R.E. (Eds)		
......	Paperback	0-947946-53-5	**£19.50**
29.	**Neurochemistry** Turner, A.J. & Bachelard, H.S. (Eds)		
......	Hardback	1-85221-028-1	**£30.00**
......	Paperback	1-85221-027-3	**£19.50**
28.	**Biological Membranes** Findlay, J.B.C. & Evans, W.H. (Eds)		
......	Hardback	0-947946-84-5	**£32.50**
......	Paperback	0-947946-83-7	**£22.50**
27.	**Nucleic Acid and Protein Sequence Analysis** Bishop, M.J. & Rawlings, C.J. (Eds)		
......	Hardback	1-85221-007-9	**£35.00**
......	Paperback	1-85221-006-0	**£25.00**
26.	**Electron Microscopy in Molecular Biology** Sommerville, J. & Scheer, U. (Eds)		
......	Hardback	0-947946-64-0	**£30.00**
......	Paperback	0-947946-54-3	**£19.50**
25.	**Teratocarcinomas and Embryonic Stem Cells** Robertson, E.J. (Ed)		
......	Hardback	1-85221-005-2	**£19.50**
......	Paperback	1-85221-004-4	**£19.50**
24.	**Spectrophotometry and Spectrofluorimetry** Harris, D.A. & Bashford, C.L. (Eds)		
......	Hardback	0-947946-69-1	**£30.00**
......	Paperback	0-947946-46-2	**£18.50**
23.	**Plasmids** Hardy, K.G. (Ed)		
......	Paperback	0-947946-81-0	**£18.50**
22.	**Biochemical Toxicology** Snell, K. & Mullock, B. (Eds)		
......	Paperback	0-947946-52-7	**£19.50**
19.	**Drosophila** Roberts, D.B. (Ed)		
......	Hardback	0-947946-66-7	**£32.50**
......	Paperback	0-947946-45-4	**£22.50**
17.	**Photosynthesis: Energy Transduction** Hipkins, M.F. & Baker, N.R. (Eds)		
......	Hardback	0-947946-63-2	**£30.00**
......	Paperback	0-947946-51-9	**£18.50**
16.	**Human Genetic Diseases** Davies, K.E. (Ed)		
......	Hardback	0-947946-76-4	**£30.00**
......	Paperback	0-947946-75-6	**£18.50**

14.	**Nucleic Acid Hybridisation** Hames, B.D. & Higgins, S.J. (Eds)		
......	Hardback	0-947946-61-6	**£30.00**
......	Paperback	0-947946-23-3	**£19.50**
13.	**Immobilised Cells and Enzymes** Woodward, J. (Ed)		
......	Hardback	0-947946-60-8	**£18.50**
12.	**Plant Cell Culture** Dixon, R.A. (Ed)		
......	Paperback	0-947946-22-5	**£19.50**
11a.	**DNA Cloning: Volume I** Glover, D.M. (Ed)		
......	Paperback	0-947946-18-7	**£18.50**
11b.	**DNA Cloning: Volume II** Glover, D.M. (Ed)		
......	Paperback	0-947946-19-5	**£19.50**
10.	**Virology** Mahy, B.W.J. (Ed)		
......	Paperback	0-904147-78-9	**£19.50**
9.	**Affinity Chromatography** Dean, P.D.G., Johnson, W.S. & others (Eds)		
......	Paperback	0-904147-71-1	**£19.50**
7.	**Microcomputers in Biology** Ireland, C.R. & Long, S.P. (Eds)		
......	Paperback	0-904147-57-6	**£18.00**
6.	**Oligonucleotide Synthesis** Gait, M.J. (Ed)		
......	Paperback	0-904147-74-6	**£18.50**
5.	**Transcription and Translation** Hames, B.D. & Higgins, S.J. (Eds)		
......	Paperback	0-904147-52-5	**£22.50**
3.	**Iodinated Density Gradient Media** Rickwood, D. (Ed)		
......	Paperback	0-904147-51-7	**£19.50**

Sets

Essential Molecular Biology: Volumes I and II as a set Brown, T.A. (Ed)		
Spiralbound hardback	0-19-963114-X	**£58.00**
Paperback	0-19-963115-8	**£40.00**
Antibodies: Volumes I and II as a set Catty, D. (Ed)		
Paperback	0-19-963063-1	**£33.00**
Cellular and Molecular Neurobiology Chad, J. & Wheal, H. (Eds)		
Spiralbound hardback	0-19-963255-3	**£56.00**
Paperback	0-19-963254-5	**£38.00**
Protein Structure and Protein Function: Two-volume set Creighton, T.E. (Ed)		
Spiralbound hardback	0-19-963064-X	**£55.00**
Paperback	0-19-963065-8	**£38.00**
DNA Cloning: Volumes I, II, III as a set Glover, D.M. (Ed)		
Paperback	1-85221-069-9	**£46.00**
Molecular Plant Pathology: Volumes I and II as a set Gurr, S.J., McPherson, M.J. & others (Eds)		
Spiralbound hardback	0-19-963354-1	**£56.00**
Paperback	0-19-963353-3	**£37.00**
Protein Purification Methods, and Protein Purification Applications, two-volume set Harris, E.L.V. & Angal, S. (Eds)		
Spiralbound hardback	0-19-963048-8	**£48.00**
Paperback	0-19-963049-6	**£32.00**
Diagnostic Molecular Pathology: Volumes I and II as a set Herrington, C.S. & McGee, J. O'D. (Eds)		
Spiralbound hardback	0-19-963241-3	**£54.00**
Paperback	0-19-963240-5	**£35.00**
Receptor Biochemistry; Receptor-Effector Coupling; Receptor-Ligand Interactions Hulme, E.C. (Ed)		
Spiralbound hardback	0-19-963096-8	**£90.00**
Paperback	0-19-963097-6	**£62.50**
Signal Transduction Milligan, G. (Ed)		
Spiralbound hardback	0-19-963296-0	**£30.00**
Paperback	0-19-963295-2	**£18.50**
Human Cytogenetics: Volumes I and II as a set (2/e) Rooney, D.E. & Czepulkowski, B.H. (Eds)		
Hardback	0-19-963314-2	**£58.50**
Paperback	0-19-963313-4	**£40.50**
Peptide Hormone Secretion/Peptide Hormone Action Siddle, K. & Hutton, J.C. (Eds)		
Spiralbound hardback	0-19-963072-0	**£55.00**
Paperback	0-19-963073-9	**£38.00**

ORDER FORM for UK, Europe and Rest of World

(Excluding USA and Canada)

Qty	ISBN	Author	Title	Amount
			P&P	
			TOTAL	

Please add postage and packing: £1.75 for UK orders under £20; £2.75 for UK orders over £20; overseas orders add 10% of total.

Name ....

Address ....

....

.... Post code ....

[ ] Please charge £ .... to my credit card
Access/VISA/Eurocard/AMEX/Diners Club (circle appropriate card)

Card No .... Expiry date ....

Signature ....

Credit card account address if different from above:

....

.... Postcode ....

[ ] I enclose a cheque for £.....................

Please return this form to: OUP Distribution Services, Saxon Way West, Corby, Northants NN18 9ES

OR ORDER BY CREDIT CARD HOTLINE: Tel +44-(0)536-741519 or
Fax +44-(0)536-746337

ORDER OTHER TITLES OF INTEREST TODAY

123.	**Protein Phosphorylation** Hardie, G. (Ed)		
......	Spiralbound hardback	0-19-963306-1	**$65.00**
......	Paperback	0-19-963305-3	**$45.00**
121.	**Tumour Immunobiology** Gallagher, G., Rees, R.C. & others (Eds)		
......	Spiralbound hardback	0-19-963370-3	**$72.00**
......	Paperback	0-19-963369-X	**$50.00**
117.	**Gene Transcription** Hames, D.B. & Higgins, S.J. (Eds)		
......	Spiralbound hardback	0-19-963292-8	**$72.00**
......	Paperback	0-19-963291-X	**$50.00**
116.	**Electrophysiology** Wallis, D.I. (Ed)		
......	Spiralbound hardback	0-19-963348-7	**$66.50**
......	Paperback	0-19-963347-9	**$45.95**
115.	**Biological Data Analysis** Fry, J.C. (Ed)		
......	Spiralbound hardback	0-19-963340-1	**$80.00**
......	Paperback	0-19-963339-8	**$60.00**
114.	**Experimental Neuroanatomy** Bolam, J.P. (Ed)		
......	Spiralbound hardback	0-19-963326-6	**$65.00**
......	Paperback	0-19-963325-8	**$40.00**
111.	**Haemopoiesis** Testa, N.G. & Molineux, G. (Eds)		
......	Spiralbound hardback	0-19-963366-5	**$65.00**
......	Paperback	0-19-963365-7	**$45.00**
113.	**Preparative Centrifugation** Rickwood, D. (Ed)		
......	Spiralbound hardback	0-19-963208-1	**$90.00**
......	Paperback	0-19-963211-1	**$50.00**
110.	**Pollination Ecology** Dafni, A.		
......	Spiralbound hardback	0-19-963299-5	**$65.00**
......	Paperback	0-19-963298-7	**$45.00**
109.	**In Situ Hybridization** Wilkinson, D.G. (Ed)		
......	Spiralbound hardback	0-19-963328-2	**$58.00**
......	Paperback	0-19-963327-4	**$36.00**
108.	**Protein Engineering** Rees, A.R., Sternberg, M.J.E. & others (Eds)		
......	Spiralbound hardback	0-19-963139-5	**$75.00**
......	Paperback	0-19-963138-7	**$50.00**
107.	**Cell-Cell Interactions** Stevenson, B.R., Gallin, W.J. & others (Eds)		
......	Spiralbound hardback	0-19-963319-3	**$60.00**
......	Paperback	0-19-963318-5	**$40.00**
106.	**Diagnostic Molecular Pathology: Volume I** Herrington, C.S. & McGee, J. O'D. (Eds)		
......	Spiralbound hardback	0-19-963237-5	**$58.00**
......	Paperback	0-19-963236-7	**$38.00**
105.	**Biomechanics-Materials** Vincent, J.F.V. (Ed)		
......	Spiralbound hardback	0-19-963223-5	**$70.00**
......	Paperback	0-19-963222-7	**$50.00**
104.	**Animal Cell Culture (2/e)** Freshney, R.I. (Ed)		
......	Spiralbound hardback	0-19-963212-X	**$60.00**
......	Paperback	0-19-963213-8	**$40.00**
103.	**Molecular Plant Pathology: Volume II** Gurr, S.J., McPherson, M.J. & others (Eds)		
......	Spiralbound hardback	0-19-963352-5	**$65.00**
......	Paperback	0-19-963351-7	**$45.00**
101.	**Protein Targeting** Magee, A.I. & Wileman, T. (Eds)		
......	Spiralbound hardback	0-19-963206-5	**$75.00**
......	Paperback	0-19-963210-3	**$50.00**
100.	**Diagnostic Molecular Pathology: Volume II: Cell and Tissue Genotyping** Herrington, C.S. & McGee, J.O'D. (Eds)		
......	Spiralbound hardback	0-19-963239-1	**$60.00**
......	Paperback	0-19-963238-3	**$39.00**
99.	**Neuronal Cell Lines** Wood, J.N. (Ed)		
......	Spiralbound hardback	0-19-963346-0	**$68.00**
......	Paperback	0-19-963345-2	**$48.00**
98.	**Neural Transplantation** Dunnett, S.B. & Björklund, A. (Eds)		
......	Spiralbound hardback	0-19-963286-3	**$69.00**
......	Paperback	0-19-963285-5	**$42.00**
97.	**Human Cytogenetics: Volume II: Malignancy and Acquired Abnormalities (2/e)** Rooney, D.E. & Czepulkowski, B.H. (Eds)		
......	Spiralbound hardback	0-19-963290-1	**$75.00**
......	Paperback	0-19-963289-8	**$50.00**
96.	**Human Cytogenetics: Volume I: Constitutional Analysis (2/e)** Rooney, D.E. & Czepulkowski, B.H. (Eds)		
......	Spiralbound hardback	0-19-963288-X	**$75.00**
......	Paperback	0-19-963287-1	**$50.00**
95.	**Lipid Modification of Proteins** Hooper, N.M. & Turner, A.J. (Eds)		
......	Spiralbound hardback	0-19-963274-X	**$75.00**
......	Paperback	0-19-963273-1	**$50.00**
94.	**Biomechanics-Structures and Systems** Biewener, A.A. (Ed)		
......	Spiralbound hardback	0-19-963268-5	**$85.00**
......	Paperback	0-19-963267-7	**$50.00**
93.	**Lipoprotein Analysis** Converse, C.A. & Skinner, E.R. (Eds)		
......	Spiralbound hardback	0-19-963192-1	**$65.00**
......	Paperback	0-19-963231-6	**$42.00**
92.	**Receptor-Ligand Interactions** Hulme, E.C. (Ed)		
......	Spiralbound hardback	0-19-963090-9	**$75.00**
......	Paperback	0-19-963091-7	**$50.00**
91.	**Molecular Genetic Analysis of Populations** Hoelzel, A.R. (Ed)		
......	Spiralbound hardback	0-19-963278-2	**$65.00**
......	Paperback	0-19-963277-4	**$45.00**
90.	**Enzyme Assays** Eisenthal, R. & Danson, M.J. (Eds)		
......	Spiralbound hardback	0-19-963142-5	**$68.00**
......	Paperback	0-19-963143-3	**$48.00**
89.	**Microcomputers in Biochemistry** Bryce, C.F.A. (Ed)		
......	Spiralbound hardback	0-19-963253-7	**$60.00**
......	Paperback	0-19-963252-9	**$40.00**
88.	**The Cytoskeleton** Carraway, K.L. & Carraway, C.A.C. (Eds)		
......	Spiralbound hardback	0-19-963257-X	**$60.00**
......	Paperback	0-19-963256-1	**$40.00**
87.	**Monitoring Neuronal Activity** Stamford, J.A. (Ed)		
......	Spiralbound hardback	0-19-963244-8	**$60.00**
......	Paperback	0-19-963243-X	**$40.00**
86.	**Crystallization of Nucleic Acids and Proteins** Ducruix, A. & Gieg⟨130⟩, R. (Eds)		
......	Spiralbound hardback	0-19-963245-6	**$60.00**
......	Paperback	0-19-963246-4	**$50.00**
85.	**Molecular Plant Pathology: Volume I** Gurr, S.J., McPherson, M.J. & others (Eds)		
......	Spiralbound hardback	0-19-963103-4	**$60.00**
......	Paperback	0-19-963102-6	**$40.00**
84.	**Anaerobic Microbiology** Levett, P.N. (Ed)		
......	Spiralbound hardback	0-19-963204-9	**$75.00**
......	Paperback	0-19-963262-6	**$45.00**

83. **Oligonucleotides and Analogues** Eckstein, F. (Ed)
.... Spiralbound hardback 0-19-963280-4 **$65.00**
.... Paperback 0-19-963279-0 **$45.00**
82. **Electron Microscopy in Biology** Harris, R. (Ed)
.... Spiralbound hardback 0-19-963219-7 **$65.00**
.... Paperback 0-19-963215-4 **$45.00**
81. **Essential Molecular Biology: Volume II** Brown, T.A. (Ed)
.... Spiralbound hardback 0-19-963112-3 **$65.00**
.... Paperback 0-19-963113-1 **$45.00**
80. **Cellular Calcium** McCormack, J.G. & Cobbold, P.H. (Eds)
.... Spiralbound hardback 0-19-963131-X **$75.00**
.... Paperback 0-19-963130-1 **$50.00**
79. **Protein Architecture** Lesk, A.M.
.... Spiralbound hardback 0-19-963054-2 **$65.00**
.... Paperback 0-19-963055-0 **$45.00**
78. **Cellular Neurobiology** Chad, J. & Wheal, H. (Eds)
.... Spiralbound hardback 0-19-963106-9 **$73.00**
.... Paperback 0-19-963107-7 **$43.00**
77. **PCR** McPherson, M.J., Quirke, P. & others (Eds)
.... Spiralbound hardback 0-19-963226-X **$55.00**
.... Paperback 0-19-963196-4 **$40.00**
76. **Mammalian Cell Biotechnology** Butler, M. (Ed)
.... Spiralbound hardback 0-19-963207-3 **$60.00**
.... Paperback 0-19-963209-X **$40.00**
75. **Cytokines** Balkwill, F.R. (Ed)
.... Spiralbound hardback 0-19-963218-9 **$64.00**
.... Paperback 0-19-963214-6 **$44.00**
74. **Molecular Neurobiology** Chad, J. & Wheal, H. (Eds)
.... Spiralbound hardback 0-19-963108-5 **$56.00**
.... Paperback 0-19-963109-3 **$36.00**
73. **Directed Mutagenesis** McPherson, M.J. (Ed)
.... Spiralbound hardback 0-19-963141-7 **$55.00**
.... Paperback 0-19-963140-9 **$35.00**
72. **Essential Molecular Biology: Volume I** Brown, T.A. (Ed)
.... Spiralbound hardback 0-19-963110-7 **$65.00**
.... Paperback 0-19-963111-5 **$45.00**
71. **Peptide Hormone Action** Siddle, K. & Hutton, J.C.
.... Spiralbound hardback 0-19-963070-4 **$70.00**
.... Paperback 0-19-963071-2 **$50.00**
70. **Peptide Hormone Secretion** Hutton, J.C. & Siddle, K. (Eds)
.... Spiralbound hardback 0-19-963068-2 **$70.00**
.... Paperback 0-19-963069-0 **$50.00**
69. **Postimplantation Mammalian Embryos** Copp, A.J. & Cockroft, D.L. (Eds)
.... Spiralbound hardback 0-19-963088-7 **$70.00**
.... Paperback 0-19-963089-5 **$50.00**
68. **Receptor-Effector Coupling** Hulme, E.C. (Ed)
.... Spiralbound hardback 0-19-963094-1 **$70.00**
.... Paperback 0-19-963095-X **$45.00**
67. **Gel Electrophoresis of Proteins (2/e)** Hames, B.D. & Rickwood, D. (Eds)
.... Spiralbound hardback 0-19-963074-7 **$75.00**
.... Paperback 0-19-963075-5 **$50.00**
66. **Clinical Immunology** Gooi, H.C. & Chapel, H. (Eds)
.... Spiralbound hardback 0-19-963086-0 **$69.95**
.... Paperback 0-19-963087-9 **$50.00**
65. **Receptor Biochemistry** Hulme, E.C. (Ed)
.... Spiralbound hardback 0-19-963092-5 **$70.00**
.... Paperback 0-19-963093-3 **$50.00**
64. **Gel Electrophoresis of Nucleic Acids (2/e)** Rickwood, D. & Hames, B.D. (Eds)
.... Spiralbound hardback 0-19-963082-8 **$75.00**
.... Paperback 0-19-963083-6 **$50.00**
63. **Animal Virus Pathogenesis** Oldstone, M.B.A. (Ed)
.... Spiralbound hardback 0-19-963100-X **$68.00**
.... Paperback 0-19-963101-8 **$40.00**
62. **Flow Cytometry** Ormerod, M.G. (Ed)
.... Paperback 0-19-963053-4 **$50.00**
61. **Radioisotopes in Biology** Slater, R.J. (Ed)
.... Spiralbound hardback 0-19-963080-1 **$75.00**
.... Paperback 0-19-963081-X **$45.00**
60. **Biosensors** Cass, A.E.G. (Ed)
.... Spiralbound hardback 0-19-963046-1 **$65.00**
.... Paperback 0-19-963047-X **$43.00**

59. **Ribosomes and Protein Synthesis** Spedding, G. (Ed)
.... Spiralbound hardback 0-19-963104-2 **$75.00**
.... Paperback 0-19-963105-0 **$45.00**
58. **Liposomes** New, R.R.C. (Ed)
.... Spiralbound hardback 0-19-963076-3 **$70.00**
.... Paperback 0-19-963077-1 **$45.00**
57. **Fermentation** McNeil, B. & Harvey, L.M. (Eds)
.... Spiralbound hardback 0-19-963044-5 **$65.00**
.... Paperback 0-19-963045-3 **$39.00**
56. **Protein Purification Applications** Harris, E.L.V. & Angal, S. (Eds)
.... Spiralbound hardback 0-19-963022-4 **$54.00**
.... Paperback 0-19-963023-2 **$36.00**
55. **Nucleic Acids Sequencing** Howe, C.J. & Ward, E.S. (Eds)
.... Spiralbound hardback 0-19-963056-9 **$59.00**
.... Paperback 0-19-963057-7 **$38.00**
54. **Protein Purification Methods** Harris, E.L.V. & Angal, S. (Eds)
.... Spiralbound hardback 0-19-963002-X **$60.00**
.... Paperback 0-19-963003-8 **$40.00**
53. **Solid Phase Peptide Synthesis** Atherton, E. & Sheppard, R.C.
.... Spiralbound hardback 0-19-963066-6 **$58.00**
.... Paperback 0-19-963067-4 **$39.95**
52. **Medical Bacteriology** Hawkey, P.M. & Lewis, D.A. (Eds)
.... Spiralbound hardback 0-19-963008-9 **$69.95**
.... Paperback 0-19-963009-7 **$50.00**
51. **Proteolytic Enzymes** Beynon, R.J. & Bond, J.S. (Eds)
.... Spiralbound hardback 0-19-963058-5 **$60.00**
.... Paperback 0-19-963059-3 **$39.00**
50. **Medical Mycology** Evans, E.G.V. & Richardson, M.D. (Eds)
.... Spiralbound hardback 0-19-963010-0 **$69.95**
.... Paperback 0-19-963011-9 **$50.00**
49. **Computers in Microbiology** Bryant, T.N. & Wimpenny, J.W.T. (Eds)
.... Paperback 0-19-963015-1 **$40.00**
48. **Protein Sequencing** Findlay, J.B.C. & Geisow, M.J. (Eds)
.... Spiralbound hardback 0-19-963012-7 **$56.00**
.... Paperback 0-19-963013-5 **$38.00**
47. **Cell Growth and Division** Baserga, R. (Ed)
.... Spiralbound hardback 0-19-963026-7 **$62.00**
.... Paperback 0-19-963027-5 **$38.00**
46. **Protein Function** Creighton, T.E. (Ed)
.... Spiralbound hardback 0-19-963006-2 **$65.00**
.... Paperback 0-19-963007-0 **$45.00**
45. **Protein Structure** Creighton, T.E. (Ed)
.... Spiralbound hardback 0-19-963000-3 **$65.00**
.... Paperback 0-19-963001-1 **$45.00**
44. **Antibodies: Volume II** Catty, D. (Ed)
.... Spiralbound hardback 0-19-963018-6 **$58.00**
.... Paperback 0-19-963019-4 **$39.00**
43. **HPLC of Macromolecules** Oliver, R.W.A. (Ed)
.... Spiralbound hardback 0-19-963020-8 **$54.00**
.... Paperback 0-19-963021-6 **$45.00**
42. **Light Microscopy in Biology** Lacey, A.J. (Ed)
.... Spiralbound hardback 0-19-963036-4 **$62.00**
.... Paperback 0-19-963037-2 **$38.00**
41. **Plant Molecular Biology** Shaw, C.H. (Ed)
.... Paperback 1-85221-056-7 **$38.00**
40. **Microcomputers in Physiology** Fraser, P.J. (Ed)
.... Spiralbound hardback 1-85221-129-6 **$54.00**
.... Paperback 1-85221-130-X **$36.00**
39. **Genome Analysis** Davies, K.E. (Ed)
.... Spiralbound hardback 1-85221-109-1 **$54.00**
.... Paperback 1-85221-110-5 **$36.00**
38. **Antibodies: Volume I** Catty, D. (Ed)
.... Paperback 0-947946-85-3 **$38.00**
37. **Yeast** Campbell, I. & Duffus, J.H. (Eds)
.... Paperback 0-947946-79-9 **$36.00**
36. **Mammalian Development** Monk, M. (Ed)
.... Hardback 1-85221-030-3 **$60.00**
.... Paperback 1-85221-029-X **$45.00**
35. **Lymphocytes** Klaus, G.G.B. (Ed)
.... Hardback 1-85221-018-4 **$54.00**
34. **Lymphokines and Interferons** Clemens, M.J., Morris, A.G. & others (Eds)
.... Paperback 1-85221-035-4 **$44.00**
33. **Mitochondria** Darley-Usmar, V.M., Rickwood, D. & others (Eds)
.... Hardback 1-85221-034-6 **$65.00**
.... Paperback 1-85221-033-8 **$45.00**

32.	**Prostaglandins and Related Substances** Benedetto, C., McDonald-Gibson, R.G. & others (Eds)		
......	Hardback	1-85221-032-X	**$58.00**
......	Paperback	1-85221-031-1	**$38.00**
31.	**DNA Cloning: Volume III** Glover, D.M. (Ed)		
......	Hardback	1-85221-049-4	**$56.00**
......	Paperback	1-85221-048-6	**$36.00**
30.	**Steroid Hormones** Green, B. & Leake, R.E. (Eds)		
......	Paperback	0-947946-53-5	**$40.00**
29.	**Neurochemistry** Turner, A.J. & Bachelard, H.S. (Eds)		
......	Hardback	1-85221-028-1	**$56.00**
......	Paperback	1-85221-027-3	**$36.00**
28.	**Biological Membranes** Findlay, J.B.C. & Evans, W.H. (Eds)		
......	Hardback	0-947946-84-5	**$54.00**
......	Paperback	0-947946-83-7	**$36.00**
27.	**Nucleic Acid and Protein Sequence Analysis** Bishop, M.J. & Rawlings, C.J. (Eds)		
......	Hardback	1-85221-007-9	**$66.00**
......	Paperback	1-85221-006-0	**$44.00**
26.	**Electron Microscopy in Molecular Biology** Sommerville, J. & Scheer, U. (Eds)		
......	Hardback	0-947946-64-0	**$54.00**
......	Paperback	0-947946-54-3	**$40.00**
25.	**Teratocarcinomas and Embryonic Stem Cells** Robertson, E.J. (Ed)		
......	Hardback	1-85221-005-2	**$62.00**
......	Paperback	1-85221-004-4	**$0.00**
24.	**Spectrophotometry and Spectrofluorimetry** Harris, D.A. & Bashford, C.L. (Eds)		
......	Hardback	0-947946-69-1	**$56.00**
......	Paperback	0-947946-46-2	**$39.95**
23.	**Plasmids** Hardy, K.G. (Ed)		
......	Paperback	0-947946-81-0	**$36.00**
22.	**Biochemical Toxicology** Snell, K. & Mullock, B. (Eds)		
......	Paperback	0-947946-52-7	**$40.00**
19.	**Drosophila** Roberts, D.B. (Ed)		
......	Hardback	0-947946-66-7	**$67.50**
......	Paperback	0-947946-45-4	**$46.00**
17.	**Photosynthesis: Energy Transduction** Hipkins, M.F. & Baker, N.R. (Eds)		
......	Hardback	0-947946-63-2	**$54.00**
......	Paperback	0-947946-51-9	**$36.00**
16.	**Human Genetic Diseases** Davies, K.E. (Ed)		
......	Hardback	0-947946-76-4	**$60.00**
......	Paperback	0-947946-75-6	**$34.00**
14.	**Nucleic Acid Hybridisation** Hames, B.D. & Higgins, S.J. (Eds)		
......	Hardback	0-947946-61-6	**$60.00**
......	Paperback	0-947946-23-3	**$36.00**
13.	**Immobilised Cells and Enzymes** Woodward, J. (Ed)		
......	Hardback	0-947946-60-8	**$0.00**
12.	**Plant Cell Culture** Dixon, R.A. (Ed)		
......	Paperback	0-947946-22-5	**$36.00**
11a.	**DNA Cloning: Volume I** Glover, D.M. (Ed)		
......	Paperback	0-947946-18-7	**$36.00**
11b.	**DNA Cloning: Volume II** Glover, D.M. (Ed)		
......	Paperback	0-947946-19-5	**$36.00**
10.	**Virology** Mahy, B.W.J. (Ed)		
......	Paperback	0-904147-78-9	**$40.00**
9.	**Affinity Chromatography** Dean, P.D.G., Johnson, W.S. & others (Eds)		
......	Paperback	0-904147-71-1	**$36.00**
7.	**Microcomputers in Biology** Ireland, C.R. & Long, S.P. (Eds)		
......	Paperback	0-904147-57-6	**$36.00**
6.	**Oligonucleotide Synthesis** Gait, M.J. (Ed)		
......	Paperback	0-904147-74-6	**$38.00**
5.	**Transcription and Translation** Hames, B.D. & Higgins, S.J. (Eds)		
......	Paperback	0-904147-52-5	**$38.00**
3.	**Iodinated Density Gradient Media** Rickwood, D. (Ed)		
......	Paperback	0-904147-51-7	**$36.00**

Sets

	Essential Molecular Biology: Volumes I and II as a set Brown, T.A. (Ed)		
......	Spiralbound hardback	0-19-963114-X	**$118.00**
......	Paperback	0-19-963115-8	**$78.00**
	Antibodies: Volumes I and II as a set Catty, D. (Ed)		
......	Paperback	0-19-963063-1	**$70.00**
	Cellular and Molecular Neurobiology Chad, J. & Wheal, H. (Eds)		
......	Spiralbound hardback	0-19-963255-3	**$133.00**
......	Paperback	0-19-963254-5	**$79.00**
	Protein Structure and Protein Function: Two-volume set Creighton, T.E. (Ed)		
......	Spiralbound hardback	0-19-963064-X	**$114.00**
......	Paperback	0-19-963065-8	**$80.00**
	DNA Cloning: Volumes I, II, III as a set Glover, D.M. (Ed)		
......	Paperback	1-85221-069-9	**$92.00**
	Molecular Plant Pathology: Volumes I and II as a set Gurr, S.J., McPherson, M.J. & others (Eds)		
......	Spiralbound hardback	0-19-963354-1	**$0.00**
......	Paperback	0-19-963353-3	**$0.00**
	Protein Purification Methods, and Protein Purification Applications, two-volume set Harris, E.L.V. & Angal, S. (Eds)		
......	Spiralbound hardback	0-19-963048-8	**$98.00**
......	Paperback	0-19-963049-6	**$68.00**
	Diagnostic Molecular Pathology: Volumes I and II as a set Herrington, C.S. & McGee, J. O'D. (Eds)		
......	Spiralbound hardback	0-19-963241-3	**$0.00**
......	Paperback	0-19-963240-5	**$0.00**
	Receptor Biochemistry; Receptor-Effector Coupling; Receptor-Ligand Interactions Hulme, E.C. (Ed)		
......	Spiralbound hardback	0-19-963096-8	**$193.00**
......	Paperback	0-19-963097-6	**$125.00**
	Signal Transduction Milligan, G. (Ed)		
......	Spiralbound hardback	0-19-963296-0	**$60.00**
......	Paperback	0-19-963295-2	**$38.00**
	Human Cytogenetics: Volumes I and II as a set (2/e) Rooney, D.E. & Czepulkowski, B.H. (Eds)		
......	Hardback	0-19-963314-2	**$130.00**
......	Paperback	0-19-963313-4	**$90.00**
	Peptide Hormone Secretion/Peptide Hormone Action Siddle, K. & Hutton, J.C. (Eds)		
......	Spiralbound hardback	0-19-963072-0	**$135.00**
......	Paperback	0-19-963073-9	**$90.00**

ORDER FORM for USA and Canada

Qty	ISBN	Author	Title	Amount
			S&H	
	CA and NC residents add appropriate sales tax			
			TOTAL	

Please add shipping and handling: $2.50 for first book, ($1.00 each book thereafter)

Name ....

Address ....

....

.... Zip ....

[ ] Please charge $ .... to my credit card
Mastercard/VISA/American Express (circle appropriate card)

Acct. .... Expiry date ....

Signature ....

Credit card account address if different from above:

....

.... Zip ....

[ ] I enclose a cheque for $............

Mail orders to: Order Dept. Oxford University Press, 2001 Evans Road, Cary, NC 27513